Glucosinolates in Rapeseeds: Analytical Aspects

WORLD CROPS:
PRODUCTION, UTILIZATION, DESCRIPTION

Volume 13

1. Stanton WR, Flach M, eds: SAGO. The equatorial swamp as a natural resource. 1980. ISBN 90-247-2470-8
2. Pollmer WG, Phipps RH, eds: Improvement of quality traits of maize for grain and silage use. 1980. ISBN 90-247-2289-6
3. Bond DA, ed: *Vicia faba:* Feeding value, processing and viruses. 1980. ISBN 90-247-2362-0
4. Thompson R, ed: *Vicia faba:* Physiology and breeding. 1981. ISBN 90-247-2496-1
5. Bunting ES, ed: Production and utilization of protein and oilseed crops. 1981. ISBN 90-247-2532-1
6. Hawtin G, Webb C, eds: Faba bean improvement. 1982. ISBN 90-247-2593-3
7. Margaris N, Koedam A, Vokou D, eds: Aromatic plants: Basic and applied aspects. 1982. ISBN 90-247-2720-0
8. Thompson R, Casey R, eds: Perspectives for peas and lupins as protein crops. 1983. ISBN 90-247-2792-8
9. Saxena MC, Singh KB, eds: Ascochyta blight and winter sowing of chickpeas. 1984. ISBN 90-247-2875-4
10. Hebblethwaite PD, Dawkins TCK, Heath MC, Lockwood G, eds: *Vicia faba*: Agronomy, physiology and breeding. 1985. ISBN 90-247-2964-5
11. Sørensen H, ed: Advances in the production and utilization of cruciferous crops. 1985. ISBN 90-247-3196-8
12. Lawes DA, Thomas H, eds: Proceedings of the Second International Oats Conference. 1986. ISBN 90-247-3335-9
13. Wathelet J-P, ed: Glucosinolates in Rapeseeds: Analytical Aspects. 1987. ISBN 90-247-3525-4

Glucosinolates in Rapeseeds: Analytical Aspects

Proceedings of a Seminar in the CEC Programme of Research on Plant Productivity, held in Gembloux (Belgium), 1–3 October 1986

edited by

J-P. WATHELET
Faculté des Sciences Agronomiques de l'Etat
Chaire de Chimie Générale et Organique
B 5800 Gembloux
Belgium

1987 **MARTINUS NIJHOFF PUBLISHERS**
a member of the KLUWER ACADEMIC PUBLISHERS GROUP
DORDRECHT / BOSTON / LANCASTER

for the Commission of the European Communities

Distributors

for the United States and Canada: Kluwer Academic Publishers, P.O. Box 358, Accord Station, Hingham, MA 02018-0358, USA
for the UK and Ireland: Kluwer Academic Publishers, MTP Press Limited, Falcon House, Queen Square, Lancaster LA1 1RN, UK
for all other countries: Kluwer Academic Publishers Group, Distribution Center, P.O. Box 322, 3300 AH Dordrecht, The Netherlands

ISBN-13: 978-94-010-8118-4 e-ISBN-13: 978-94-009-3615-7
DOI: 10.1007/978-94-009-3615-7
EUR 10781 EN

Book information

Publication arranged by: Commission of the European Communities, Directorate-General Telecommunications, Information Industries and Innovation, Luxembourg

GLUCOSINOLATES IN RAPESEEDS

Analytical Aspects

CONTENTS

PREFACE

This publication contains proceedings of a Seminar on GLUCOSINOLATES IN RAPESEEDS - Analytical Aspects, held in Gembloux (Belgium) from 1 to 3 October 1986.

The meeting was organized by request of the Commission of the European Communities in the context of the CEC Programme of Research on Plant Productivity.

The main aim of the Seminar was to contribute to the elaboration of reliable quantitative methods for glucosinolate determination in rapeseeds.

Fourty Experts from thirteen countries participated in this Seminar.

Original contributions which were considered of special importance for the subject covered by the Seminar were presented and discussed.

Thanks are due to the Chairmen, Dr. Heaney R., Dr. Biston R., Dr. Ribaillier D., Prof. Dr. Robbelen G., authors and participants in the meeting for their contributions, friendliness and cooperation.

Special thanks go to Dr. Mc Gregor (Canada) and Dr. Uppström B. (Sweden) who gave Members the benefit of their knowledge.

Finally, I would also like to thank those who have helped in organizing this Seminar : Director Lecomte R. (Centre de Recherches Agronomiques de l'Etat, Gembloux), Rector Ledent A. (Faculté des Sciences Agronomiques de l'Etat, Gembloux), Prof. Severin M., Dr. Biston R., Mrs Bock and all my colleagues.

Dr. J-P Wathelet

SPEECH OF WELCOME BY Albert LEDENT

Rector of the "Faculté des Sciences Agronomiques de l'Etat"
Gembloux - Belgium

It is a privilege and a pleasure for me to offer you, on behalf of the Faculty of Agronomy and the Research Center of Agronomy, a very warm welcome in this workshop devoted to analytical aspects of glucosinolates in rapeseeds.

I would like to thank you all for honouring this meeting with your presence.

The present workshop has been organized under the auspices and with the financial support of the Commission for European Communities. We express our gratitude to this important organization for the confidence it has placed in us on this occasion.

The next few days will provide the opportunity for Specialists in our Faculty and in the Research Center to compare their results with those recorded in other European laboratories also engaged in a programme to work out a method for analysing glucosinolates present in rapeseeds.

It is known that these substances are toxic for mono and polygastric animals.

At a time when the European Economic Community is encouraging farmers to improve the quality of their production and has imposed, to that effect, much higher standards, it is essential to define and establish reproducible, precise analytical methods which clearly specify, according to accepted criteria, the quality of the products.

Considering the professional standing of the participants and the high level of papers presented, I have no doubt that the aim of your work will be attained. By the exchange of knowledge and information, the communication of your results and the final discussion, you will contribute to solving the principal problems related to analyses which I know are very complex.

Wishing you great success, I congratulate right away the organizers of this meeting.

And I also hope that, with the contribution of a generous autumn sun, you will spend a very pleasant few days at the Faculty which comprises the splendid Benedictine abbey of Gembloux.

THE BCR PROGRAMME - FOOD ANALYSIS

P.J. Wagstaffe

BCR, Commission of the European Communities,
200 rue de la Loi,1049 Brussels, Belgium

ABSTRACT

The Community Bureau of Reference (BCR) is a department of the Commission of the European Communities having the broad aim of improving the accuracy and comparability of measurements considered important at European level.For essentially chemical measurements, these aims are realised through the organisation of carefully designed intercomparaisons involving specialised laboratories and, when necessary, lead to the preparation of Reference Materials having accurately certified properties.

In the past four years, the BCR has undertaken a systematic programme in the field of food analysis. This has led to the preparation of a series of reference materials for the analysis of toxic and nutritional elements in milk*, meats*, single cell protein and cereals; mycotoxins in milk*, animal fields and cereals; composition of edible oil and fats (fatty acids and sterols); residues of illegal or controlled compounds in farm animals (stilbenes, thyreostatics) and so on.

The projects are run in close collaboration with specialists from the laboratories of the Member States, who assist in the definition of the projects, the preparation of the materials and who execute the analytical measurements. The developpement of realistic, stable and homogeneous materials presents many practical problems. Details of the programme on food are presented and the difficulties and successes illustrated by reference to examples taken from projects in the above-mentioned areas.

* Certified and available.

INTRODUCTION

The sector of the BCR programme concerned with agro-food is responding to the recognition by industrial, public health and nutritional laboratories that reliable measurement of many of the most important properties cannot be achieved through reliance on written standards alone.

Requirements of GLP, the demands imposed by laboratory accreditation and the increased awareness of the needs for in-house quality assurance schemes have caused laboratories to seek means by which they can check the reliability of their own measurements by external comparison. Such comparison is commonly made by participation in ring-tests or, more conveniently, by use of certified reference materials (CRMs), when available.

The enormity of the problem of satisfying the needs of those concerned with agro-food measurement is well reflected by the fact that there are at least 4000 individually documented methods of measurements in this field alone.

Faced with the range and complexity of analytical problems, it is unrealistic to expect all labs to be good in all fields covered by them. But, need some labs be so glad ?

Fig. 1 presents results from several recent international quality assurance programmes for organic and inorganic analytes and show the difficulty that many laboratories have with relatively common assays.

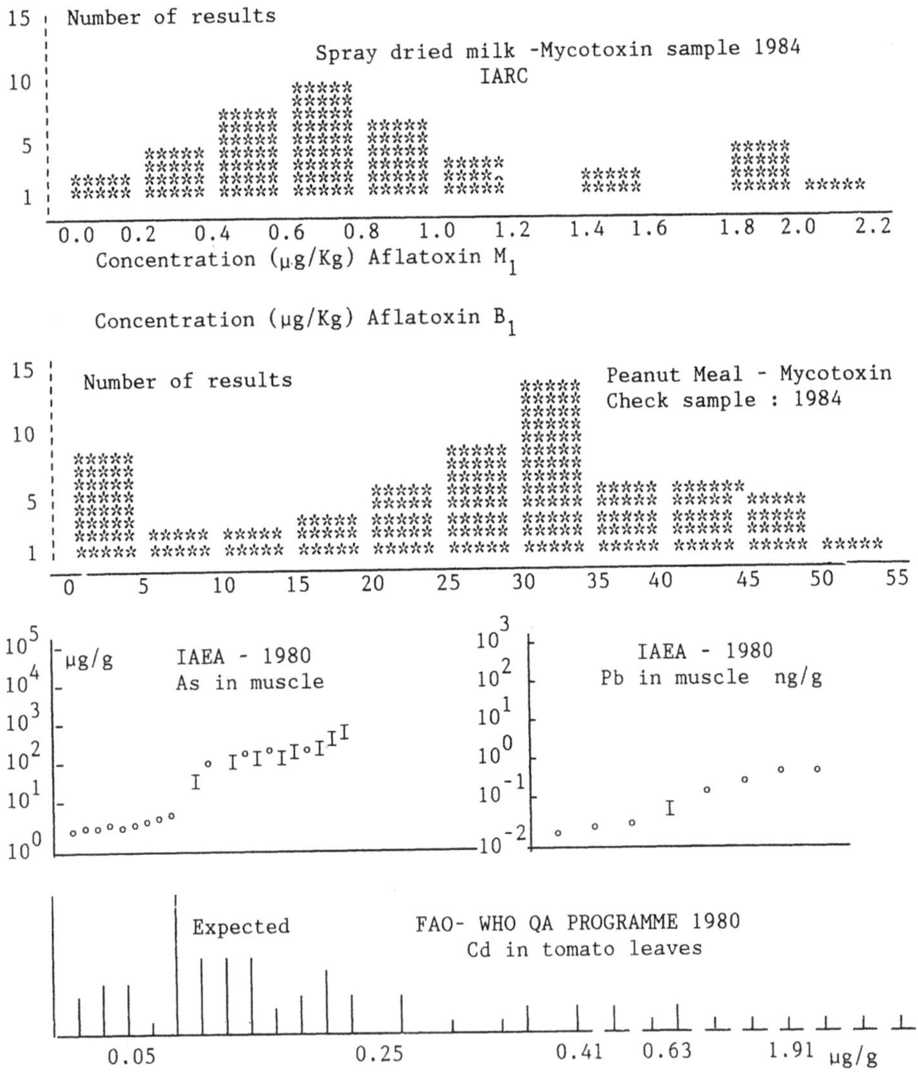

Fig. 1 Results of various international studies for determination of contaminants in food.

Whilst it is probable that most of the outlying laboratories had taken some steps to control the precision or repeatability of their methods, the range of results again show that accuracy is harder both to achieve and demonstrate.

OBJECTIVES OF THE BCR PROGRAMME

The objective of the programme is to improve the comparability and accuracy of measurements through the organisation of carefully designed intercomparaisons involving experienced European laboratories. The intercomparisons are designed both to establish the "state of the art" for the particular measurement and to identify the principal sources of error, be they due to an inherent weakness in a method or to its poor application.

The intercomparisons are repeated, modified as necessary, until the major sources of error are reduced to a level where between-laboratory and between-method agreement is acceptable.

When appropriate, the exercise may lead to the establishment of a certified reference material, the decision depending upon the expected demand from within the Community, the nature of the analytical problem and, of course, the practicality and expense of production.

CURRENT ACTIVITIES IN THE AGRO-FOOD AREA

Some of the activities currently underway (or recently completed in the agro-food area are summarised in table 1.

TABLE 1 Examples of Current and Recent Projects concerned with Agro-Food Measurement

TOPIC	NATURE OF ACTIVITY	STATUS
Toxic & Nutritional Elements		
Milk	Completed	3 CRMs available
Meat	Completed	3 CRMs available
Cereals	Int. & CRM	CRMs avail. end 86
Single cell protein	Int. & CRM	CRMs avail. end 86
Trace Organic Analysis		
Aflatoxin M_1 in milk powder	Completed	3 CRMs available
Aflatoxin B_1 in animal feed	Int. & CRM	on-going
Don & other mycotoxins in agro-products	Int.	on-going
PCBs in fish oil	Int.	on-going
Pesticides in milk	Int.	on-going
Growth promoting hormones	Int. & CRMs	on-going

Nutritional/Composition		
Fatty acids in edible oil & fats (incl. milk-fat)	Int. & CRMs	CRMs avail. end 86
Major components	Int.	on-going
Fruit juice analysis	Int.	on-going
Physical Properties		
Farinograph	Int. & CRMs	on-going
Water activity	CRM	available 1987

N.B. : Int. = Intercomparison exercises

During the past three years, 19 intercomparisons have been carried out in this sector alone, involving the collaboration of some 120 laboratories, and leading to the establishment of 8 CRMs certified for approximately 60 individual properties.

REQUIREMENTS FOR A CERTIFIED REFERENCE MATERIAL

To be satisfactory, a CRM must fulfil certain basic requirements, which are broadly described below :

Firstly, the RM must be representative of the samples actually analysed; they must be present in its normally encountered form. Thus, for example, contaminants must not be added artificially by simple "spiking".

Secondly, both the analytes and matrix must be stable and, for practical purposes, BCR aims at a life at least five years and, preferably, longer. It is thus essential to choose appropriate packaging and to demonstrate stability experimentally.

Thirdly, the reference material must be sufficiently homogeneous for the intended purpose. In the case of meat RMs for inorganic elements for example, this means that 30 kgs of lyophilised material must be uniform at the 200 mg level. In addition to the necessity to physically mix the materials on a relatively large scale, it is often difficult to find methods of adequate precision to demonstrate that the homogeneity is acceptable at the required level. For inorganic elements, multi-element techniques such as ICP and NAA are very useful; organic compounds on the other hand, often present severe difficulties in this regard.

Fourthly, the reference material must be economic and practical in usage. The final price must not be so high as to discourage its use in the routine laboratory.

Finally, and most important, the certified value must be as close to the 'true' value as practically possible, which means that it must be associated with an acceptably small uncertainty. Elimination of method-dependent errors is ensured by certifying properties on the basis of several independent methods whenever this is possible. When independent methods, performed in different laboratories give closely agreeing results, we can be sufficiently confident in the final certified value.

SOME PRACTICAL EXAMPLES FROM RECENT STUDIES

Inorganic Analyses

The objective has been to prepare a series of materials covering the principal food categories currently available or 3 milk CRMs (2 for trace elements, Cd, Pb, Hg, Cu and Zn and 1 for major elements Ca, Mg,K,N and Cl); a single cel protein (major elements) and 3 meat materials, bovine muscle, bovine liver and pig kidney certified for elements of toxicological and nutritional importance. Two cereals (wholemeal flour and brown bread) and a second single cell protein will be certified this year. Fish, vegetables and edible oil materials will be developped over the coming years.

The problems of developping such CRMs are well illustrated by the meat project, initially undertaken in support of Community legislation intended to control harmful substances in meats.

In the case of the toxic elements Cd,Pb,Hg and As, it has proved difficult to define convenient methods of adequate precision and accuracy, which are sufficiently robust to satisfy the needs for official reference methods.

Particular care was taken in selection of the materials used in each stage of preparation, i.e. collection of 70 kg wet meats, cutting with titanium knives, lyophilisation, grinding in an agate mill, sieving,mixing in a high-grade polyethylene drum and, finally, bottling.

A preliminary intercomparison of methods for a range of important elements was organised amongst experienced veterinary and food laboratories and included several laboratories specialised in the application of sophisticated methods, such as INAA and IDMS.

The conclusions of this study are summarised in the form of ranges of results in table 2.

Detailed discussion of the results with the participants revealed a number of sources of error including : losses or contamination during sample preparation, especially where poorly controlled dry-ashing was used, application of instrumental methods below the limits of determination, (e.g. Flame

AAS and ICP for Pb and Cd), inadequate preliminary treatment before final determination (e.g. incomplete oxidation of organic matter with ASV methods, incomplete reduction of Se VI to Se IV before hybride generation for Se determination).

TABLE 2 Results (Ranges) of the preliminary intercomparison of methods on a bovine liver material. (The ranges of results obtained in the final certification exercice for bovine liver RM 185 are given for comparison.)
p = number of sets of results received

	Preliminary Intercomparison		Certification Exercise	
	Range of mean values	p	Range of mean values	p
Cd ng/g	200 - 1290	22	257 - 333	16
Pb ng/g	200 - 1400	22	276 - 400	17
Hg ng/g	25 - 220	6	33 72	9
As ng/g	7 - 30	5	23 - 50	7
Se ng/g	150 - 610	7	423 - 540	12
Cu ng/g	159 - 393	21	171 - 199	16
Zn ug/g	114 - 162	27	153 - 181	20
Fe ug/g	148 - 278	24	193 - 230	18
Mn ug/g	5 - 13	17	8.3 - 10.7	13

For the final reference materials, bovine muscle, bovine liver and pig kidney were selected since it is well known that many elements accumulate preferentially in liver and kidney, whereas the levels in muscle tissue are relatively low. Thus, these 3 materials would provide a convenient range of concentration for many elements of interest, as well as giving some variation in matrix and "between-element" effects.

The certification exercise was carried out by a group of 28 laboratories (5 to 17 per element) and was designed so as to both eliminate problems encountered in the first study and to ensure maximum variety in methods of sample digestion and final determination. Laboratories were required to make 5 independent determinations of each element, paying special attention to choice of calibrant, calibration, use of methods that offered sufficient sensitivity and accuracy at the expected concentration and, application of a dry-weight correction as determined by a specified method.

The certified values and the corresponding uncertainties for the elemental contents of the three materials are summarised in table 3.

TABLE 3 Certified Elemental Contents and Uncertainties in RMs 184,185 and 186

Element	CRM 184 Bovine Muscle			CRM 185 Bovine Liver			CRM 186 Pig Kidney		
	Certified Value	Uncertainty	p	Certified Value	Uncertainty	p	Certified Value	Uncertainty	p
Cd	13 ng/g	± 2 ng/g	12	298 ng/g	± 25 ng/g*	13	2.71 μg/g	± 0.15 μg/g*	15
Pb	239 ng/g	± 11 ng/g	11	501 ng/g	± 27 ng/g	10	306 ng/g	± 11 ng/g	12
Hg	2.6 ng/g	± 0.6 ng/g	5	44 ng/g	± 3 ng/g	6	1.97 μg/g	± 0.04 μg/g	8
As				24 ng/g	± 3 ng/g	5	63 ng/g	± 9 ng/g	5
Se	183 ng/g	± 12 ng/g	12	466 ng/g	± 25 ng/g	12	10.5 μg/g	± 0.6 μg/g	11
Cu	2.3 ùg/g	± 0.06 μg/g	10	189 μg/g	± 4 μg/g	12	31.9 μg/g	± 0.4 μg/g	10
Zn	166 μg/g	± 3 μg/g	17	142 μg/g	± 3 ùg/g	17	128 μg/g	± 3 μg/g	16
Fe	79 μg/g	± 2 μg/g	15	214 μg/g	± 5 μg/g	14	299 μg/g	± 9 μg/g	14
Mn	334 ng/g	± 28 ng/g	6	9.3 μg/g	± 0.2 μg/g	9	8.4 μg/g	± 0.3 μg/g	10

It is of interest to note the wide range of elemental contents in the three RMs, particularly for Cd and Hg and, to a lesser extent, for Se, Cu and Mn.

The results accepted for certification of Cd and Hg are presented in Fig. 2 in the form bar-charts.

Indicative values were also established for the major elements Na, K, Cl, Mg, Ca and P, generally based on measurements in three laboratories using a variety of methods. Data are also provided for I (by two laboratories using NAA with radio nuclear separation) and for Cr and Ni.

Trace Organics Mycotoxins

A number of BCR intercomparison and RM projects have been undertaken for PACs, PCBs and mycotoxins. The problems ef ensuring that the certified values are accurate are however in general more difficult than for trace elements. These problems are demonstrated by the series of mycotoxin projects, summarised in table 4.

The programme began with a feasibility study carried out in a number of laboratories, designed to establish whether in each case, it was actually possible to prepare homogeneous and stable materials that would fulfil the essential requirements of a CRM.

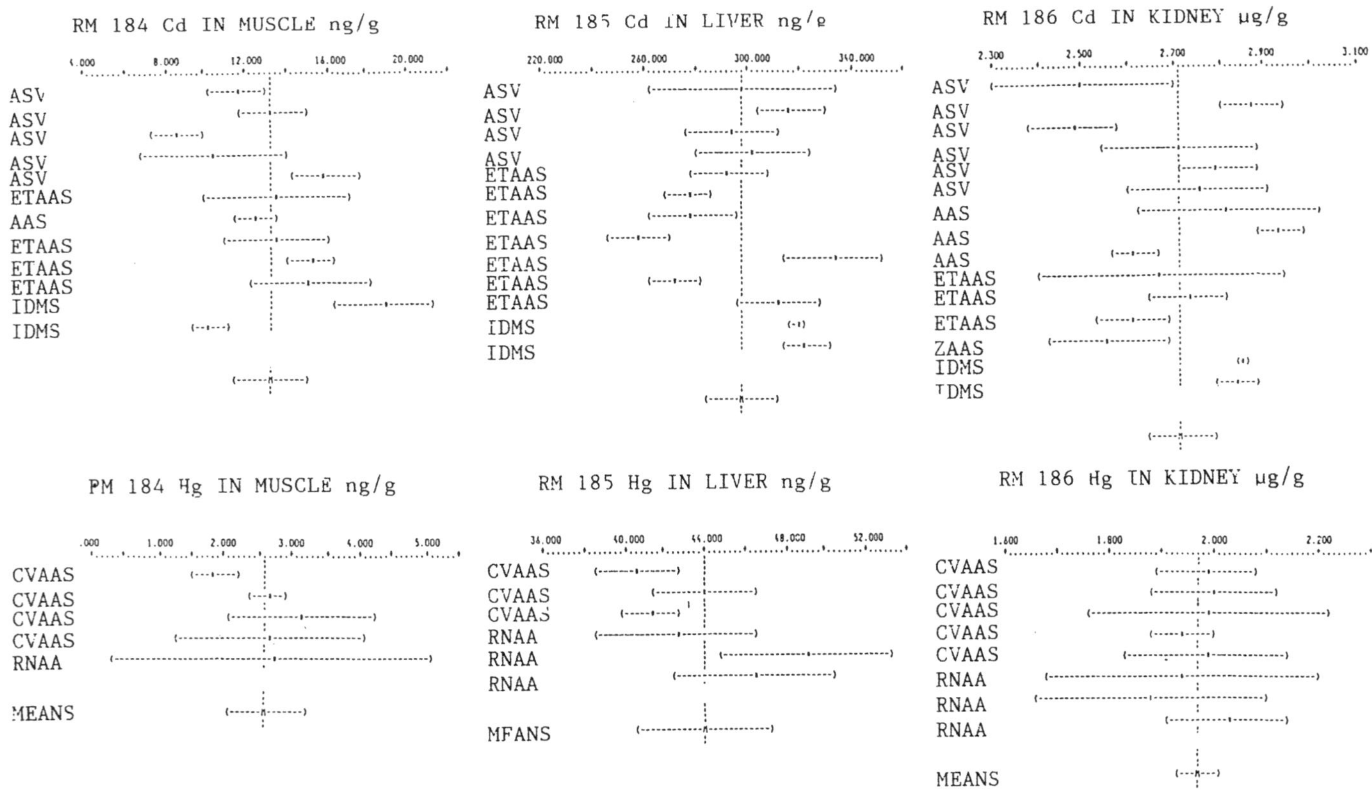

Fig. 2 Bar charts of results accepted for the certification of Cd and Hg in CRMs 184, 185 and 186 (Abbreviations for methods of final determination are as recommended by IUPAC).

TABLE 4 Summary of BCR Mycotoxin Matrix RM Projects

Matrix :	Milk powder	Penaut Meal + Compound Feeds	Wheat
Mycotoxin :	Aflatoxin M_1 Certified .05;.31 & .76 ug/kg	Aflatoxin B_1 10-40 ug/kg	Deoxynivalenol 400 ug/kg
Preparation :	Spray-drying Milk of cattle Fed Aflatoxin B_1	"naturally" Encountered Products	Natural & Fungal induced Contamination
Note : Zearalenone and ochratoxin A will be undertaken in a 3rd phase.			

Considerable emphasis was placed on ensuring that the analytes were present in the natural form. Thus, the aflatoxin M_1 in milk powder material was prepared from milk obtained from cows fed on a diet containing controlled amounts of aflatoxin B_1 .

The milk-powder was subjected to extensive homogeneity and stability studies, the latter being carried out at 4 temperatures over a 12 month period with no evidence of degradation in aflatoxin M_1, even at +47°C.

In parallel with these studies, a series of intercomparison was carried out involving some 20 experienced European Laboratories.

As you can seen from Fig. 3a, the results varied from 0.1 to 1.3 ug/kg Participants had also determined the aflatoxin M_1 content of a given simple chloroform solution using their own calibrants. The fact that the results for this solution were almost as scattered as for the milk powder suggested serious problems with the individual calibrants.

A second study was organised which was designed to allow the major sources of error to be identified. Participants were provided with : a common calibrant solution; an unknown plain chloroform solution of aflatoxin M_1; an ampouled chloroformic milk extract and a milk powder sample.

The results of the second study (Fig. 3b) were a little better but, more important, by examining the results for the simple chloroform solution and milk extracts, it was possible to demonstrate that many of the high and low results were inaccurate.

This allowed design of the final certificaiton exercises. In this final study, participants were asked to make 8 independent determinations to the best of their ability employing the method that they considered to be most important in their hands. In addition, they were provided with an ampouled calibrant solution of aflatoxin M_1 in chloroform which was prepared by dilution of a primary calibrant which had been checked by 3 independent laboratories according to the AOAC procedure.

As is evident from Fig. 3c, very satisfactory agreement was achieved in the certification exercise.

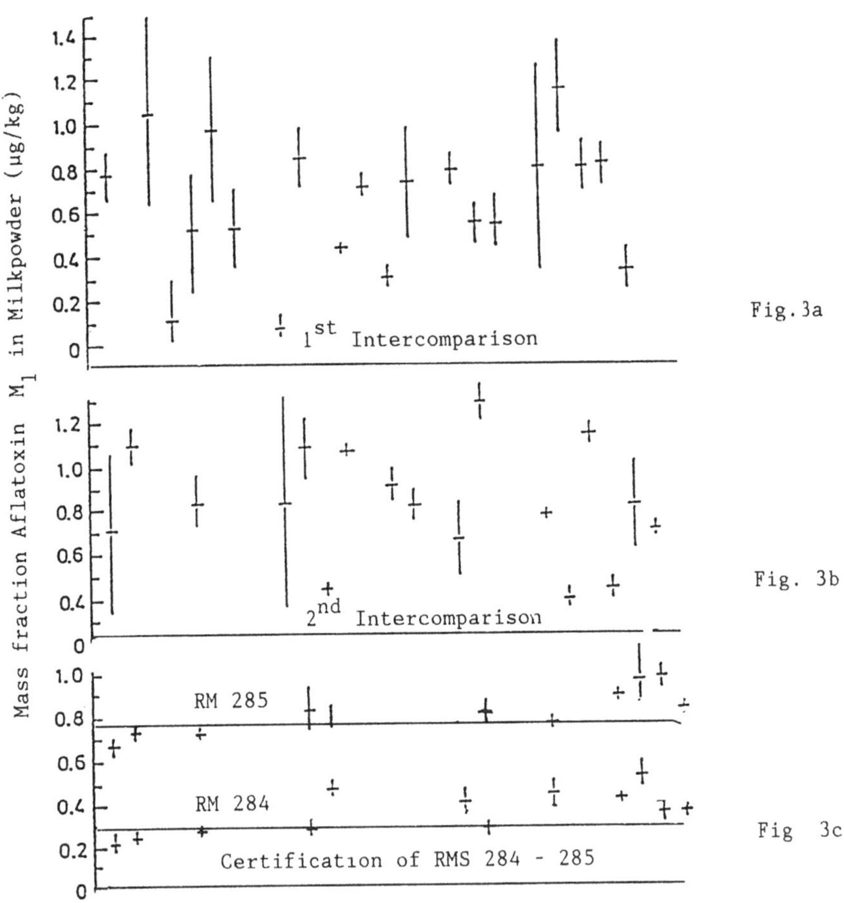

Fig. 3a, 3b and 3c Comparison of results of preliminary intercomparison with those of the certification exercise.

The 3 CRMs, which were prepared from naturally contaminated milk, have been subjected to exhaustive homogeneity and stability tests. They are available in units of 25 g with following certified values :

	Aflatoxin M_1 content
CRM 282	0.05 μg (an effective blank)
CRM 284	0.31 +/- 0.06 μg/Kg
CRM 285	0.76 +/- 0.05 μg/Kg

Développement of the aflatoxin B_1 peanut meal material has followed a path similar to that for M_1.

The first intercomparison of methods, involving 22 generally experienced laboratories, was performed on homogenized batches of peanut meal and peanut butter (circa 100 μg/Kg).

In addition, a common calibrant consisting of a chloroform solution of aflatoxin B_1, B_2,G_1 and G_2 was provided as well as an "unknown" mixture in chloroform to check calibration and manipulative aspects.

The results of this study for aflatoxin B_1 in peanut meal are presented in Fig. 4.

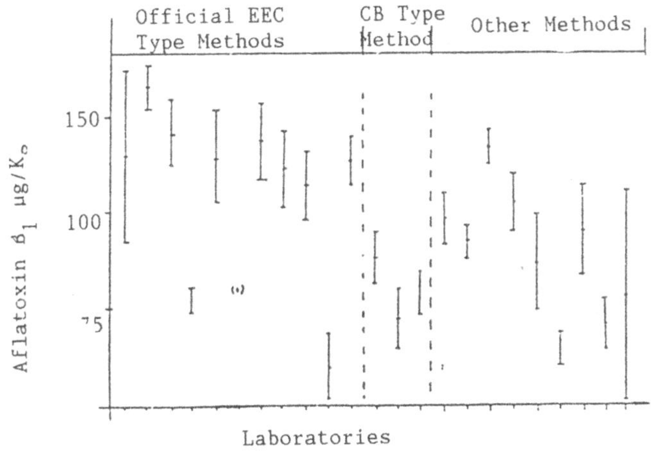

Fig. 4 1st Intercomparison of methods for Aflatoxin B_1 in peanut meal

It is evident from Fig. 4 that even at the rather high level of circa 120 μg/Kg, there is poor agreement for aflatoxin B_1 in peanut meal. Although there was reasonably good agreement between results obtained by the EEC official method there was no means of telling where the true value lay.

It was concluded that considerable improvement was necessary before certification of a peanut meal could be achieved.

A 2nd study was therefore organized but restricted to determination of aflatoxin B_1 in peanut meal at circa 50 μg/Kg.

The study was divided into 4 stages : - calibration - calibrant supplied

- unmoulded meal : recovery experiments
- detoxified meal extract : clean-up efficiency spiked with aflatoxin B_1
- normal peanut meal : Overall performance at circa 50 μg/Kg

Fig. 5 summarises the results for aflatoxin B_1 after correction for recovery (as determined in parallel spiking experiments). Although detailed technical evaluation with the participants is yet to be done, it is evident that a very high level of agreement has been achieved. This is a most satisfactory conclusion especially considering the variety of methods used and the results of earlier BCR and other studies (cf IARC study 1984, Fig.1).

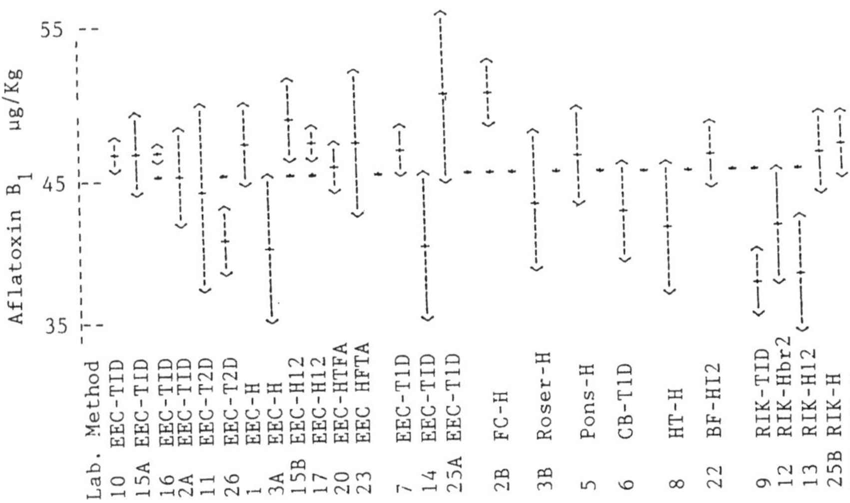

Fig. 5 Aflatoxin B_1 in peanut meal - 2nd Study (Corected for recovery)

The BCR group is now in a strong position to undertake final certification and to move to lower concentrations.

FUTURE WORK

There is little difficulty in identifying areas where there is a need to improve the accuracy and comparability of results obtained in food analysis. The major problem is that of selecting priorities. Table 5 presents some of the topics which have been proposed as being the most important and where we propose to work in the next few years.

TABLE 5 Future RM Projects for Food Analysis

Inorganic analysis	Organic Contaminants	General
Edible oil	PAC, PCB,	Vitamins
	Sea food toxins,	Sugars
Vegetable	Prohibited hormones	Dietary Fibre
Fish	and other compounds	Wine
	in Farm Animals	Fruit juice
	Pesticides ...	
Food Technology :	Water sorption isotherm	
	Farinograph calibration	

Your comments on these proposals and suggestion for other areas which you consider to be of high priority will be most welcome.

We would also be pleased to hear from you if you would like to collaborate in projects where you have a particular interest.

ACKNOWLEDGEMENTS

The skill and support of the participating laboratories, on whom the success on individual projects depends, is warmly acknowledged.

COMPARISON OF MYROSINASE ACTIVITY ASSAYS

Sandro Palmieri, Renato Iori, Onofrio Leoni
Istituto Sperimentale per le Colture Industriali
Via di Corticella,133 40129 Bologna, Italy.

ABSTRACT

Plant thioglucoside glucohydrolase (myrosinase EC 3.2.3.1) catalyzes the hydrolysis of glucosinolates to isothiocyanate, glucose and sulfate. We have purified myrosinase from white mustard seed (*Sinapis alba*) in high yields and with a considerable specific activity in a single step by affinity chromatography on Con-A Sepharose. This myrosinase was suitable for use in our polarographic method, which allows the simultaneous determination of total glucosinolates and free glucose in cruciferous extracts. We compared four methods for measuring myrosinase activity for linearity, sensitivity, reproducibility and suitability for routine work, both for routine analyses of crude myrosinase extracts and for enzyme purification and characterization studies. The methods were: i) pH-Stat Assay (pHSA), which measures the rate of acid released during the hydrolysis of sinigrin substrate; ii) Spectrophotometric Coupled Enzyme Assay (SCEA), employing hexokinase and glucose-6-phosphate dehydrogenase (HK--G6PDH), which measures the rate of glucose released during sinigrin hydrolysis; iii) Direct Spectrophotometric Assay (DSA), in which the substrate hydrolysis rate was measured by following the decrease in absorbance at 227 nm; and, finally, iv) a new Polarographic Coupled Assay (PCA), involving the coupled enzymes glucose oxidase and catalase, which measures the rate of glucose released during substrate hydrolysis as O_2 uptake. Although pHSA and PCA showed comparable activities and were linear with increasing amounts of purified enzyme greater than 10 µg, these methods are not suitable for routine work because pHSA requires extensive sample dialysis and PCA shows poor sensitivity. DSA and SCEA gave 18% and 33% lower myrosinase activities, respectively, compared to pHSA; in addition, SCEA was nonlinear with increasing amounts of enzyme above 1 µg. As expected, the lower activity for DSA was due to the suboptimum substrate concentration while for SCEA, this result depends to the low concentrations of Mg^{++} and HK--G6PDH. In conclusion, DSA appears to be better than the other methods both for enzyme kinetic studies and, if used with care, for routine analyses of crude myrosinase.

INTRODUCTION

Plant myrosinase (thioglucoside glucohydrolase, EC 3.2.3.1) is widespread in seeds of the family Cruciferae and catalyzes the hydrolysis of glucosinolates, also contained in cruciferous seeds, to form goitrogenic and potentially hepatoxic isothiocyanates, glucose, and sulfate.

The importance of myrosinase for glucosinolate analyses is evident because it is well established that many of these techniques require myrosinase for glucosinolate hydrolysis before analysis. One can also see the importance of glucosinolate analyses for commercial utilization of the defatted meals as animal feed, for their potential technological utilization and, finally, for screening in breeding programs. In recent years several papers have reported useful analytical techniques suitable for total and individual glucosinolate determination in cruciferous material; these are UV spectrophotometry (Wetter and Youngs, 1976), gas-chromatography (Underhill and Kirkland,1971; Thies, 1976), tititrimetry (Croft, 1979), ion-exchange chromatography (Heaney and Fenwick, 1981) and HPLC (Helboe et al., 1980).

In a previous paper, (Iori et al. 1983) we described a method to determine total glucosinolates and endogenous glucose content simultaneously by polarographic determination of O_2 uptake using the coupled enzymes glucose-oxidase--myrosinase. As reported in Fig.1, the method affords the free glucose of the sample in the same analysis time (ca.4min) of the total glucosinolates, allowing the processing of a large number of samples per day, starting from the crude extracts. In addition, this method allows the analysis of samples with low glucosinolate contents (< 10 μmol/g). On the basis of our experience with more than 1500 samples of rapeseed analyzed for glucosinolate and glucose in a national project for yield and quality improvement of oleaginous plants supported by the Italian Ministry of Agriculture, we can report that the technique gives good results for quality of data, cost, and analysis time. Nevertheless, the applicability of the polarographic method greatly depends on the availability of myrosinase having good specific activity.

After several purification experiments, carried out with typical multicolum systems (Palmieri et al. 1982), we purified myrosinase from white mustard seeds (Sinapis alba), starting from aqueous dialyzed extract in a single step by affinity chromatography on Con A-Sepharose (Fig.2) obtaining high specific activity (over 20 000 U/mg) and good recovery (over 90%) (Palmieri et al., 1986). Myrosinase isolated by this simple procedure migrated on SDS-polyacrylamide gel as a nearly homogeneous polypeptide with a molecular weight of about 140 000 as shown in Figure 3.

Myrosinase, besides being essential in many procedures for glucosinolate analyses, appears to be an important enzyme for the biological and technological implications in the food and feed quality and the safety of cruciferous material. It is clear that a dependable, rapid and inexpensive assay to measure myrosinase activity is essential for studies in this field.

Over the last decade,numerous publications have described indirect and direct techniques for assaying myrosinase from several sources. The direct methods, viz. titration of released acid with alkali using pH-stat apparatus (Björkman and Lönnerdal, 1973), spectrophotometric measurements of the decrease in absorbance at 227nm during sinigrin disappearance (Schwimmer,1961; Gil and MacLeod, 1980; Palmieri et al., 1982) and,recently, a spectrophotometric coupled assay that measures the glucose released via hexokinase--glucose-6-phosphate dehydrogenase (Wilkinson et al.,1984a-b), appear to be better than the indirect ones and are in theory all equally efficient, simple to use and suitable for kinetic analyses as they allow one to determine myrosinase activity continuously. However, to our knowledge the relative advantages and disadvantages of the cited methods have never been extensively and objectively examined, especially in view of their utilization in breeding programs and in comparative studies of different plant tissues and cruciferous species.

Our studies on cruciferous oil-bearing seeds prompted our interest in finding other convenient methods for routine myrosinase determination that overcome the potential differences between enzyme sources and/or isoenzymes that could affect activity measurement. Therefore, the purpose of this study was to find a simple, dependable and inexpensive assay that can rapidly analyze large numbers of crude myrosinase samples. With this aim an additional new polarographic assay, not previously used for this purpose, was tested. The present paper is a comparative study of four methods and also examines the effect of ascorbic acid.

MATERIALS AND METHODS

Materials. Commercial varieties of rape seeds, *Brassica napus* L. cv. Jet Neuf and cv. Jade (low glucosinolate content), were obtained from Ringot (Lille, France) and Norddeutsche Pflanzenzucht (Hohenliet,W.Germany) respectively.White mustard seeds, *Sinapis alba* L. cv. Albatros, was supplied by S.I.S. Foraggera (Bologna, Italy).

Preparation of Partially Purified Myrosinase Extracts. The defatted meals were homogenized with distilled water (1:20 w/v). The insoluble material was removed by centrifugation at 17 700 g for 20 minutes, the supernatants were filtered with filter paper, dialyzed thoroughly against distilled water and the precipitated material was removed by centrifugation. The volume of the partial purified myrosinase extracts was measured and then analyzed for myrosinase activity and protein content.

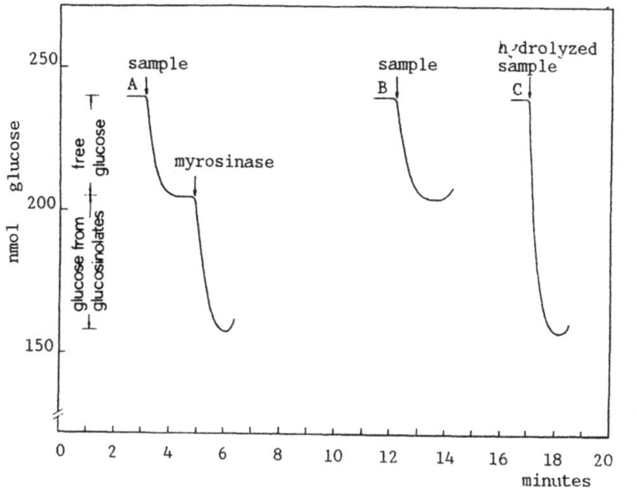

Fig.1 Polarographic and enzymatic determination of free glucose and total glucosinolates in a sample of rapeseed defatted meal with a simultaneous (A) and discontinuous (B-C) methods.

Myrosinase Purification. The enzyme extracted from white mustard was purified according to our previously reported procedure (Palmieri et al., 1986), lyophilized and stored in small amounts at -20 ° C.

Direct Spectrophotometric Assay (DSA). The activity was determined by measuring the decomposition of the sinigrin substrate by following the decrease in absorbance at 227 nm using quartz cells with a 0.5 cm path-length in a mod.219 Cary recording spectrophotometer. To compare this method with the other assays,the standard reaction mixtures and the conditions were slightly modified from those previously described (Palmieri et al., 1982) as follows. The temperature was 30 °C (although the best signal-to-noise ratio has been determined to be at 37°C, which is particularly advantageous for crude extracts analyses). The standard reaction mixture contained 0.5 m$\underline{M}$ sinigrin, 33.1 m$\underline{M}$ phosphate buffer pH 6.5 and 100 µl of appropriately diluted myrosinase in a total volume of 1.5 ml.

Spectrophotometric Enzyme Coupled Assay (SCEA). The conditions and the standard mixture were essentially those reported by Wilkinson et al. (1984a). The activity was measured at 30°C observing the increase of absorbance at 340 nm during the formation of NADPH from NADP using 1 cm path-length quartz cells in a mod. 219 Cary recording spectrophotometer.

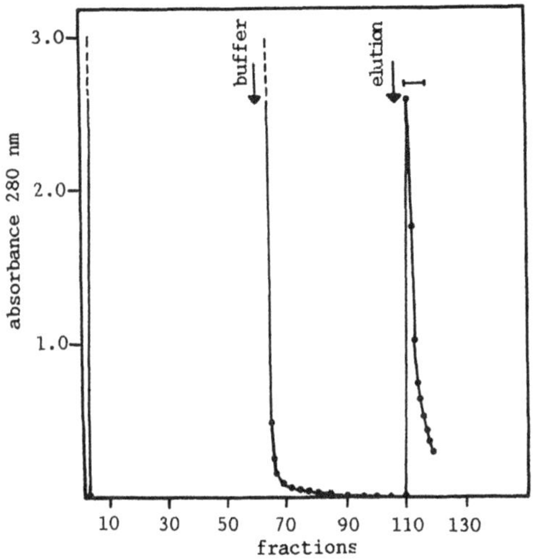

Fig.2 Dialyzed-centrifuged extract (90ml) applied to 1x10cm column containing 1ml of Con A-Sepharose at a rate of 8 ml/h. The column was then washed with starting buffer. The fraction size was 1.5ml. The eluition with 0.25 M methyl α-D-mannoside in starting buffer was begun with fraction 108; the flow was stopped for 30 min after the collection of each 1.25 ml/fraction. The enzyme was pooled as indicated.

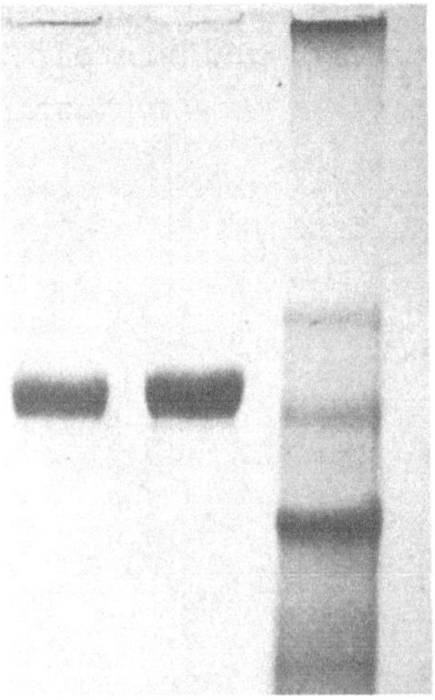

Fig.3 SDS-PAGE of myrosinase purified in a single step by Con A-Sepharose affinity chromatography. Samples were run in 7% gel, stained with Coomassie Brilliant Blue R-250. Key: myrosinase eluted with methylα-D-glucoside(left), myrosinase eluted with methyl α-D-mannoside (middle), molecular weight standards (right).

pH-Stat Assay (pHSA). The composition of the standard mixture was 5 mM sinigrin 81 mM NaCl and 50 µl appropriately diluted myrosinase and 0.5 or 1 mM ascorbate when present. The total assay volume before titration was 5 ml and all reagents, myrosinase included, were prepared in triple quartz-distilled water. The pH was adjusted to 6.5. Myrosinase activity was determined by measuring the acid release rate by titrating with 1 mM NaOH with a E 457 Metrohom dispenser (Herisau, Switzerland) connected to a 5 ml EA 928-5 burette, and a mod. PHM63 Radiometer pH-meter (Copenhagen) maintaining the pH at 6.5. During the reaction the solution was kept at 30 °C in a termostatically controlled cell and gently stirred magnetically.

Polarographic Coupled Assay (PCA). The myrosinase activity test was performed by determining the rate of oxygen uptake with a Clark electrode in a Gilson K-IC Oxygraph, Medical Electronics (Middleton, Wisc.). An excess of glucose oxidase was used to oxidize quickly the released glucose produced from the glucosinolate hydrolysis catalyzed by myrosinase according to the following general reaction:

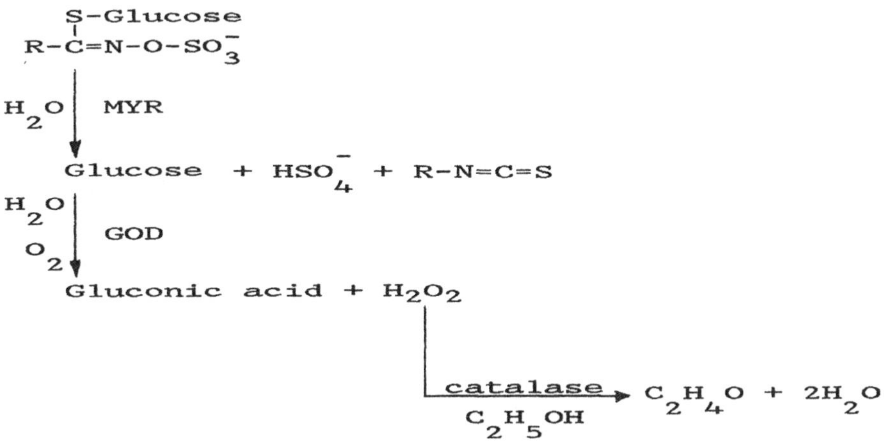

The glucose oxidase solution had the same composition as that used in a previous study for determining glucosinolate content in cruciferous material (Iori et al., 1983). The reaction mixture for determining myrosinase activity contained the glucose oxidase solution, 5 m*M* sinigrin and appropriately diluted myrosinase. During the reaction it was termostatically maintained at 30 °C and gently stirred magnetically in a total volume of 1 ml. Before use, the glucose oxidase solution was maintained at 30 °C for at least 2 h and stirred occasionally.

Protein Determination. Protein concentration was measured by the method described by Bradford (1976) using the Bio-Rad Protein Assay.

RESULTS AND DISCUSSION

In the present study the specific activity data and the ranges of linearity are not comparable to those of previous papers on this subject since the myrosinase used, although improved in purity, was in lyophilized form.

As reported by other workers (Wilkinson et al., 1984) pHSA should be considered as the reference method, although in its use some care must be taken in choosing the alkali concentration to achieve measurable reaction rates.

<u>Assay linearity, sensitivity and reproducibility</u>. Fig.4 shows the effect of enzyme concentration on the activity of the myrosinase-catalyzed reaction of sinigrin hydrolysis as measured by the four methods. pHSA and PCA show good linearity in a very wide range of enzyme concentration (Fig.4A) (presumably beyond 10 µg), whereas the range of linearity for DSA and SCEA is much narrower (Fig.4B).

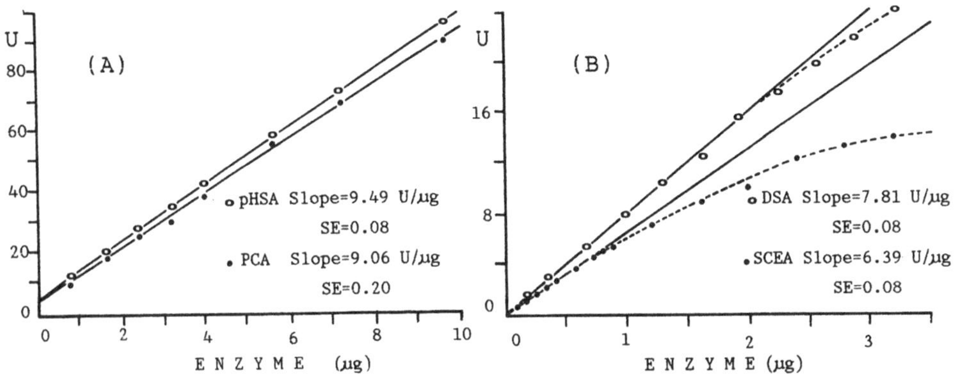

Fig.4 Enzyme concentration dependence of myrosinase activity: (A) pHSA and PCA; (B) DSA and SCEA. The specific activities calculated from the slope of each curve in the rage of linearity are also reported.

In fact, the activity measured with SCEA is linearly proportional to the amount of myrosinase only up to 0.8 µg; that obtained by DSA appears to be linear until about 2 µg enzyme. The values of specific activity are reported next to the curves in Fig.4. These were obtained by calculating the slope of each curve in its range of linearity. These data clearly show that the activity measured by PCA is the closest to that determined by pHSA, and the slight difference does not appear significant (P=0.05, df 12). On the other hand, DSA and SCEA show activities which are about 18% and 33% smaller than pHSA, respectively. The difference recorded for DSA was expected because of the lower substrate concentration (0.5 m<u>M</u>), which is near the maximum for a cell with a 0.5 cm path-length. However, this substrate concentration is about three times higher than the apparent Km value (0.156 mM (Palmieri et al., 1982), 0.17 mM (Björkman and Lönnerdal, 1973)) and therefore high enough to determine the initial velocity of the hydrolysis reaction for steady-state kinetic studies of enzyme. On the other hand, the low activity obtained by SCEA is surprising, since in this case the substrate concentration (5 m<u>M</u>) was the same as that used for pHSA and PCA. This loss of activity presumably depends on the low global first-order rate constant of the second part of reaction that transforms glucose to 6-phosphogluconate, which is catalyzed by hexokinase (HK) and glucose-6-phosphate dehydrogenase (G6PDH) with a consequent glucose accumulation during the reaction.

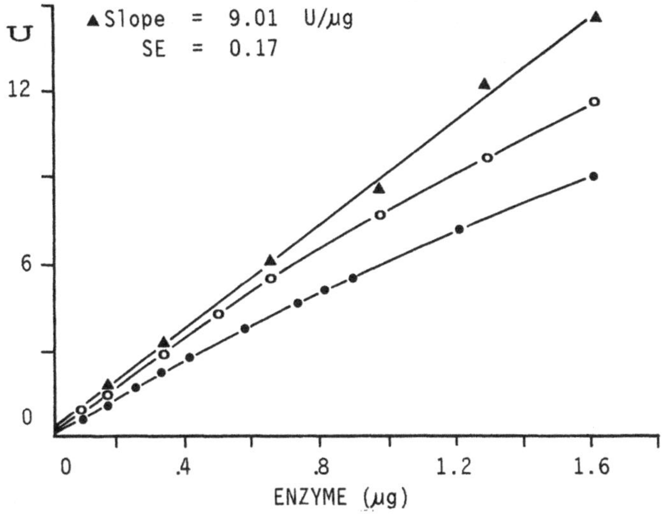

Fig.5 Enzyme concentration dependence of myrosinase activity determined by SCEA in the presence of different amounts of Mg^{++}, HK and G6PDH. (●) Mg^{++} 3mM, HK 0.56 U, G6PDH 0.35 U; (o) Mg^{++} 30 mM, HK 0.56 U; G6PDH 0.35 U; (▲) Mg^{++} 30mM, HK 5.6 U, G6PDH 3.5 U.

We can now consider the reasons for the lower activity obtained by SCEA. The results reported in Fig.5 demonstrate that the concentrations of $MgCl_2$ and coupled enzymes (HK and G6PDH) proposed by previous workers (Wilkinson et al., 1984a) are insufficient to obtain a linearity and sensitivity comparable to the other assays. In fact, as Fig.5 shows, as one increases the concentrations of Mg ion and coupled enzymes (HK and G6PDH) the resulting activities become proportional to the myrosinase concentration used, thus restoring the linear response of the assay. In these conditions the specific activity of myrosinase determined by SCEA was not significantly different from that measured by pHSA and PCA (P=0.05, df 10).

Table 1 shows the specific myrosinase activity in the crude extracts obtained from two varieties of rapeseed with high and low glucosinolate contents (Jet Neuf and Jade, respectively) and from white mustard seed determined by the four methods of analysis compared.

In all cases the activity measured by pHSA appears higher than the other assays, thus providing good sensitivity for analyzing crude extracts with low activities. If one considers the data obtained with DSA and SCEA, those for rapeseed are in good agreement, while in the data for white mustard extract, in which the activity is much higher, SCEA shows a considerable loss of sensitivity, thus confirming the results achieved for pure myrosinase. The polarographic method, as mentioned above, proved to be unsuitable for measuring low activities, since it loses sensitivity and reproducibility at these levels of activity.

TABLE 1 Specific myrosinase activity in partially purified extracts of white mustard and two rapeseed varieties.

SAMPLES	DSA (U/mg)	SCEA (U/mg)	pHSA (U/mg)	PCA (U/mg)
Rapeseed (cv.Jet Neuf)	229 ± 2	230 ± 2	264 ± 2	n.d.
Rapeseed (cv.Jade)	139 ± 1	142 ± 1	160 ± 3	n.d.
White mustard	3580 ± 15	2565 ± 15	4155 ± 35	4140±30

Effect of ascorbate. It does not appear convenient to use ascorbate as activator in the assay mixture to measure the activity of myrosinases from different sources as been proposed for SCEA (Wilkinson et al. 1984ab). In fact, Björkman and Lönnerdal (1973) demonstrated that activation with ascorbate strongly depends on the plant species, and, moreover, varieties within the species also differ in this property. In addition, contrary to a previous paper (Wilkinson et al., 1984a) which reported the same activity trend for pHSA and SCEA, the specific activities that we measured for the four assays in the presence of 0.5 m*M* ascorbate as activator differed remarkably (Fig.6A and B).

The methods considered are probably inadequate when used in the presence of this activator. For example, in the case of PCA, the catalase used in the assay mixture could be, at least in part inhibited by ascorbate (Orr, 1967), which can also act as a scavenger of the free radicals involved in the reaction chain of glucose oxidation. The reduced linearity observed for SCEA (Fig.4B) becomes even worse when ascorbate is used (Fig. 6B), thus making it difficult to choose the assay conditions and, in this case, to calculate the slope for the specific activities. The use of ascorbate with DSA and pHSA within the known limits of concentration seems more

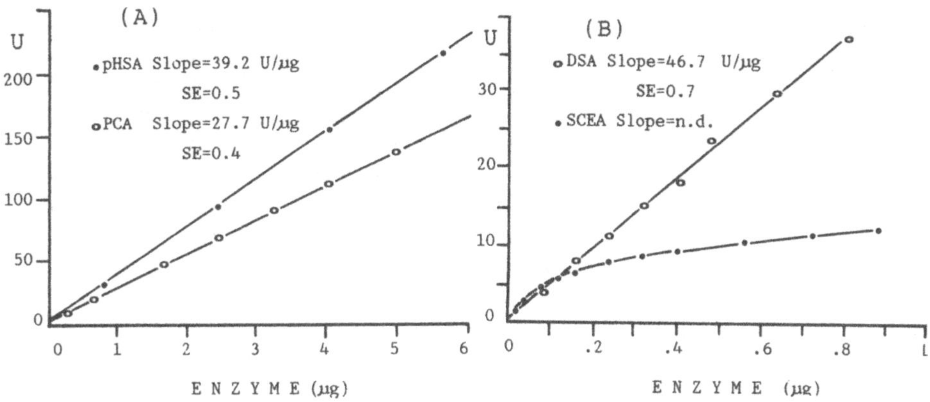

Fig.6 Enzyme concentration dependence of myrosinase activity in the presence of 0.5 mM ascorbate as enzyme activator: (A)pHSA and PCA; (B) DSA and SCEA. The specific activities calculated from the slope of each curve in the range of linearity are also reported.

feasible than the other methods, even if from the comparison of their activity data it appears that the value obtained with DSA is significantly higher. This result could be attributed to the different [enzyme] / [activator] ratio used in the two assays. In conclusion, since the effect of chemical and physical conditions on the activation mechanism of myrosinase by ascorbate is still poorly understood, we are convinced that the use of this activator in myrosinase assays should be avoided, particularly when the reaction mixture is complex, as are PCA and SCEA.

CONCLUSIONS

In addition to the results concerning the accuracy, precision and applicability of the enzyme concentration range in the compared methods, some considerations on their suitability seem in order both for routine analyses and for enzyme purification and characterization studies. Although pHSA is the most reliable assay, it has a number of drawbacks when used in routine work. The main disadvantage is without doubt the need to dialyze thoroughly all buffered myrosinase extracts before analysis, which greatly increases the analysis time. Although in the conditions used PCA had a wide range of linearity, it shows poor sensitivity. This is why we feel polarography cannot be used for routine myrosinase activity measurements in crude rapeseed extracts, even if it is a useful alternative technique to measure the activity during the enzyme purification process starting from white mustard. In theory, SCEA appears to have good characteristics for use in routine work. However, this method, as originally proposed by the authors, i.e. in the presence of ascorbate as activator (Wilkinson et al.,1984a), gave results that disagree with the other assays tested. In addition, these results are clearly affected by the non-linearity of the method, as shown in Fig.6B. This aspect was pointed out, but not investigated, by Wilkinson et al. (1984b) in a paper in which myrosinase was studied in several cruciferous vegetables. On the other hand, we think that SCEA would be suitable for routine analyses of crude extracts if it is employed without ascorbate and if the concentrations of Mg ion and coupled enzymes (HK-G6PDH) are increased in the reaction mixture at least 10 times with respect to the original method. We feel it our duty, however, to point out that the proposed variations would involve an increase in the analysis cost, which is already rather high compared to the other assays. With regard to the applicability of this assay to enzyme characterization studies, it shows some not negligible drawbacks attributable mainly to the particular molecular properties of HK and G6PDH, enzymes that are much more delicate and sensitive to variations in pH and temperature than myrosinase. For kinetic studies we believe that the DSA, as shown in a previous work (Palmieri et al., 1982), remains the most suitable assay, both for its simplicity of execution and the reliability of its results, especially when a high-performance spectrophotometer in good working conditions is available. In fact, the best feature of this technique is the possibility of observing the enzyme reaction rate continuously and directly, with the advantage that one can use a wide range of temperatures, pHs and possibly, within certain limits of concentration, some effector compounds. In addition, we think that DSA can also be useful in routine analyses of crude extracts, even though the signal-to-noise ratio must be improved for each sample with low activity by decreasing the high absorbance values of the extracts by dialysis or rapid gel-filtration systems.

REFERENCES

Björkman,R. 1976. Properties and function of plant myrosinases,in: The Biology and Chemistry of Cruciferae, p.191-205 Academic Press, London.

Björkman,R.,Lönnerdal,B. 1973. Studies on myrosinases. III. Enzymatic properties of myrosinases from Sinapis alba and Brassica napus seeds. Biochim. Biophys. Acta, 327, 121-131.

Bradford,M. 1976. A rapid and sensitive method for the quantitation of microgram quantities of protein utilizing the principle of protein-dye binding. Anal. Biochem., 72,248-254.

Croft,A.G. 1979. The determination of total glucosinolates in rapeseed meal by titration of enzyme-liberated acid and the identification of individual glucosinolates. J.Sci.Food Agric., 30,417-423.

Gil,V., MacLeod, A.J. 1980. Studies on glucosinolate degradation in Lepidium Sativum seed extracts. Phytochemistry, 19, 2547-2551.

Heaney R.K.,Fenwick G.R. 1981. A micro-column method for the rapid determination of the total glucosinolate content of cruciferous material. Z. Planzenzucht. 87,89-95.

Iori,R.,Leoni,O.,Palmieri,S. 1983. A polarographic method for the simultaneous determination of total glucosinolates and free glucose of cruciferous material. Anal.Biochem., 134, 195-198.

Helboe,P.,Olsen,O.,Sorensen,H. 1980. Separation of glucosinolates by high performance liquid chromatography. J.Chromatography, 197, 199-205.

Orr,C.W.M. 1967. Studies on Ascorbic acid. I. Factors influencing the ascorbate-mediated inhibition of catalase. Biochemistry, 6, 2995-3000.

Palmieri,S.,Leoni,O.,Iori,R. 1982. A steady-state study of myrosinase with direct ultraviolet spectrophotometric assay. Anal.Biochem., 123, 320-324.

Palmieri,S.,Iori,R.,Leoni,O. 1986. Myrosinase from Sinapis alba L.: A new method of purification for glucosinolate analyses. J.Agric.Food Chem., 34,138-140

Schwimmer, S. 1961. Spectral changes during the action of myrosinase on sinigrin. Acta Chem. Scand., 15, 535-544.

Thies,W. 1976. Quantitative gas-liquid chromatography of glucosinolates on a microliter scale. Fette Seifen, Anstrichm. 1976, 78, 231-234.

Tookey,H.L.,Van Etten,C.H.,Daxenbichler,M.E. 1980. Glucosinolates. in: Toxic Constituents of Plant Foodstuffs 2ed p.103-142. Academic Press,New York.

Underhill,E.W.and Kirkland, D.F. 1971. Gas chromatography of trimethylsilyl derivatives of glucosinolates, J.Chromatogr., 57,47

Van Etten, C.H., Daxenbichler,M.E., Peters,J.E., Booth A.N. 1965. J.Agric.Food Chem., 13/1, 24-36.

Wetter, L.R., Youngs, C.G. 1976. A thiourea-UV assay for total glucosinolate content in rapeseed meals. J.Am.Oil Chem.Soc.,53, 162.

Wilkinson,A.P.,Rhodes,M.J.C.,Fenwick,G.R. 1984a. Determination of myrosinase activity by a spectrophotometric coupled enzyme assay. Anal.Biochem., 139, 284-291.

Wilkinson,A.P.,Rhodes,M.J.C.,Fenwick,G.R. 1984b. Myrosinase activity of cruciferous vegetables. J.Sci.Food Agric., 35, 543-552.

OPTIMIZATION OF SILYLATION REACTIONS OF DESULPHOGLUCOSINOLATES BEFORE GAS CHROMATOGRAPHY

A.Landerouin, A. Quinsac, D. Ribaillier

Centre Technique Interprofessionnel des Oléagineux métropolitains (CETIOM)
Laboratoire d'Analyses - Avenue de la Pomme de Pin
Ardon - 45160-Olivet, France.

ABSTRACT

Gas chromatography offers a certain interest for the determination of glucosinolates in rapeseed seeds and meals because of its simplicity and reduced costs.

However, it is necessary to know its reliability to determine indolglucosinolates and more particularly 4-OH-glucobrassicin , for which the silylation-stage appears to be critical.

Such reasons implied the study of the two most commonly used silylation-reagents : pyridine reagent and methyl,1-imidazole reagent. The behaviour variations of the different desulphoglucosinolates were considered according to the reaction-length and temperature.

The obtained results allowed us to conclude that under certain well-defined conditions, the methyl,1-imidazole reagent was most appropriate for a routine determination of the TMS-derivatives in desulphoglucosinolates.

INTRODUCTION

Presently, the analysis of glucosinolates is carried out following two main techniques :

- the gas chromatography of desulphoglucosinolates either under isothermal conditions (THIES 1980, 1983), or with a temperature-programming (HEANEY and FENWICK, 1980),
- the high performance liquid chromatography (HPLC) either of desulphoglucosinolates (SPINKS and col., 1983), or of intact glucosinolates (HELBOE and col. 1980).

The isothermal gas chromatography does not give the possibility to determine indolglucosinolates, the relative quantity of which is more important in low-glucosinolates rapeseed varieties. This determination is possible if we use the gas chromatography with a temperature-programming, or the high performance liquid chromatography (HPLC) of desulpho-glucosinolates or intact glucosinolates.

The HPLC is probably the technique to be chosen to determine all the glucosinolates individually, but it requires a sophisticated equipment, highly developed technical means and a considerable operational budget.

At a time when the European Community is about to choose one or several method(s) to analyse the Community rapeseed production, it is useful to examine whether the gas chromatography may be a reliable method. As a matter of fact, it only requires an equipment which is already existing in most laboratories and reduced operational costs.

The determination methods of total glucosinolates by glucose-determination (RUGRAFF and al.1986),or by measurement of the complex formed with palladium (MOLLER and al.) must be studied too. To this end, we were led to precise the possibilities of gas chromatography for the determination of indolglucosinolates and more particularly of 4-hydroxy-glucobrassicin (4-OH-GBS). It is probable that the critical phase is the time of silylation-reactions. That is why we tried to reach their optimization and studied the action of the different intervening parameters (nature of the silylating agent, temperature and duration of the derivatization).

MATERIAL

Commonly used lab-equipment, and among others :

- a micro-screw-crusher (type Ernst Schutt-Göttingen, Western Germany),
- an analytical balance,
- 5ml-polypropylene tubes,
- a water-bath adjustable at 95 + 2°C,
- a timer,
- an agitator with Vortex-effect,
- a centrifuge giving the possibility to obtain an acceleration of 1500 g,
- 150 mm-long Pasteur pipettes,

- glass wool,
- 1 ml- polystyrene-tubes,
- a disposo-tray,
- a warm-air fan (hair-drier),
- reaction tubes in 1 or 2 ml-thick borosilicated glass with a screw-stopper and 'PTFE' joint,
- aluminium paper,
- two heating chambers, one adjusted at 80 + 2°C, and the second one at 120 + 2°C,
- desiccator with a deshydrating-agent (P2O5),
- a gas chromatograph with a temperature-programming equipped with a 1,8m-long column in borosilicated glass, inside diameter of 2mm, filled with 2% of OV7 on chromosorb W-AW DMCS (150-180 µm),
- an integrator-recorder.

PRODUCTS

- demineralized water, or with a rather equivalent purity
- 1 mmol/l sinigrin solution (allylglucosinolate),
- barium acetate and lead acetate, 0,5 mol/l solution,
- pyridin acetate, 3 M,
- pyridin acetate, 0,02 M,
- sulphatase, type H1 from Helix pomatia (prealably purified, controlled and diluted to 1/5)(QUINSAC and RIBAILLIER, 1984),
- saccharose 1 mM,
- pure methanol,
- stearic acid in the form of methyl ester 10 mM in acetone,
- silylation-reagent A (according to THIES)
  - . pyridin 10 V
  - . N-methyl-N-trimethylsilylheptafluorobutyramid (MSHFBA) 10 V
  - . trimethylchlorosilan (TMCS) 1 V
- silylation-reagent B (according to HEANEY)
  - . methyl-1-imidazol 1 V } 5 V
  - . acetone 19 V }

- MSHFBA 10 V
- TMCS 1 V

WORKING METHOD

Glucosinolates Extraction :

- weigh 0,2 g of rapeseed seeds,
- put them into the water bath at 95°C for 5 min.,
- add 1 ml of boiling water, stir with the agitator Vortex,
- let in the water bath for 5 min.,
- add 1 ml of sinigrin-solution to 1 mmol/l used as internal standard. Stir and let in the water bath for five more minutes.
- let cool,
- add 100 µl of the barium acetate and lead acetate solution,
- centrifuge for 5 min. (1500 g).

In order to have a sufficient quantity of extract for the comparative studies, the experiment was carried out on 3 0,2 g-samples.

Desulphation :

- deposit 500 µl of supernatent on each column (previously reactivated by pyridin acetate 3 M). Two deposits are made from a seed-sample of 0,2 g.
- wash each column with twice 1 ml of pyridin acetate 0,02 M,
- deposit 50 µl of sulphatase diluted to 1/5 on each column,
- let act at room temperature for one night,
- eluate each column with 500 µl of water, recover the effluent into a 1 ml-polypropylene tube,
- add 100 µl of the saccharose solution to each tube. Stir vigorously.
- place 200 µl of the mixture into a cup of the disposo-tray (3 cups are prepared from a tube),
- dry off with a hair-drier,

- retake the residue with 30 µl of methanol,
- bring the 18 methanolic extracts together (3 weighings, x 2 deposits x 3 cups),
- add the silylation-control (methylester of stearic acid) in a quantity rather equivalent to that of sinigrin, i.e. 150 µl,
- stir thoroughly.

Silylation :
- deposit 20 µl of the methanolic mixture into each silylation-flask,
- oven-evaporate at 80°C for 30 min.,
- let cool in the desiccator awaiting silylation,
- each tube is treated individually for the silylation. After adding 20 µl of the silylating reagent, it is closed hermetically, placing a piece of aluminium sheet between the tube and the stopper (with 'PTFE' joint). Then, it is brought to the oven for the time and at the temperature under study.

Chromatography :
- 0,5 µl of the preparation are taken out through the aluminium sheet and injected into a chromatograph GIRDEL 3000 equipped with a 1,8 m-long column, inside diameter of 2 mm, filled with chromosorb WHP impregnated with 2% of OV7 ;
- conditions of chromatography :
  . temperature of the injector : 280°C,
  . temperature of the detector : 290°C,
  . temperature of the oven : programmation from 185°C to 280°C, at the rate of 10°C per minute,
  . vector gas : hydrogene with a flow of 40 ml/min.,
  . flame ionization detector.

RESULTS AND DISCUSSION

In this study, we have compared :
- The 2 silylating reagents (mentioned above),

- 3 temperatures of silylation : 25, 80 and 120°C,
- and silylation periods from 5 to 100 min.

In order to compare the different results, all the experiments were carried out following a precise chronology :

. 1st day : extraction and desulphation,

. 2nd day : elution and evaporation,

. 3rd day : silylation and chromatography.

The figure 1 presents a characteristic chromatogram with the peaks of the different glucosinolates for the desulphation control (saccharose) and the silylation control (methylic ester of stearic acid C 18).

It was possible to compare the different glucosinolates thanks to the presence of the silylation control.

The relation : $\frac{\text{peak area of the glucosinolate}}{\text{peak area C18}}$

was calculated for each major aliphatic glucosinolate (the sinigrin control included) as well as for indole glucosinolates (GBS and 4-OH-GBS).

Influence of the silylation period :

We note (Fig. 2 and 3) that the silylation-period varies according to the used reagent. With FENWICK's reagent, a step is reached after 10 minutes at room temperature. On the other hand, with THIES's reagent, under the same conditions, 40 min. are necessary. This period may be reduced to 20 minutes if the reaction takes place at 120°C.

Influence of temperature :

With FENWICK's reagent, the temperature of silylation has little effect ; at room temperature, from the 3rd minute, the maximum ratio is reached. At 80°C, the quantity of silylated glucosinolates is roughly greater, but the difference with other temperatures is not significant. At 120°C, however, a slight increase is observed in the case of longer periods of time (74 min.),but the absence of in-between temperatures makes it difficult to interpret the results.

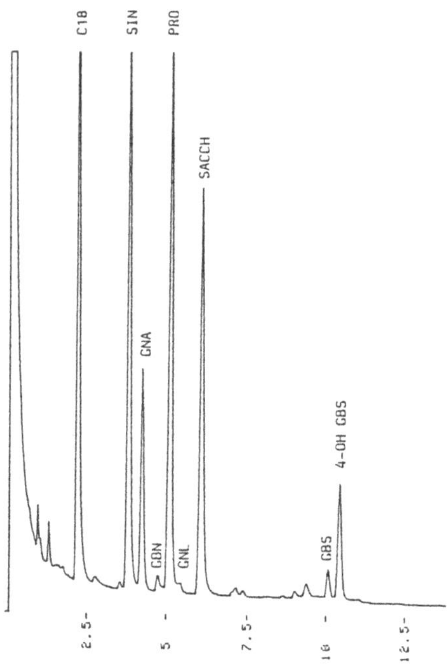

FIGURE 1

Silylation time	Reagent	t °C	GLS Area / C18 Area					
			SIN	GNA	PRO	SACCH	GBS	4-OH GBS
40 min (THIES)	THIES	room	0,70	0,25	0,55	0,51	0,05	0,14
		80°	0,72	0,25	0,64	0,50	0,06	0,36
		120°	0,70	0,24	0,56	0,50	0,07	0,28
10 min. (FENWICK)	FENWICK	room	0,80	0,27	0,69	0,54	0,07	0,36
		80°	0,83	0,29	0,73	0,56	0,09	0,43
		120°	0,79	0,28	0,71	0,54	0,08	0,45

TABLEAU 1

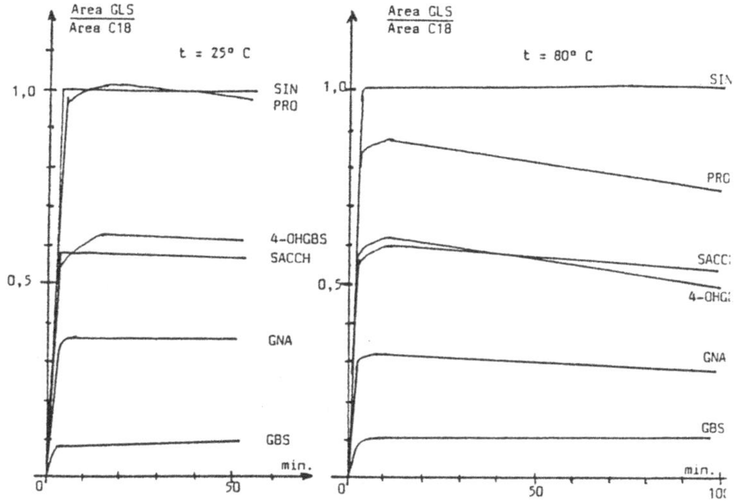

FIGURE 2

SILYLATION WITH
FENWICK REAGENT

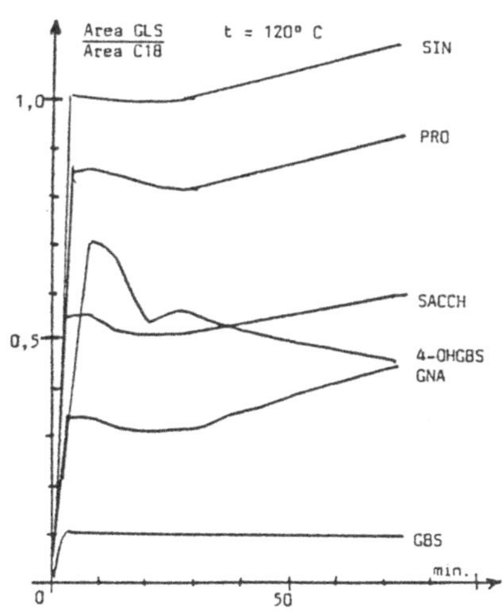

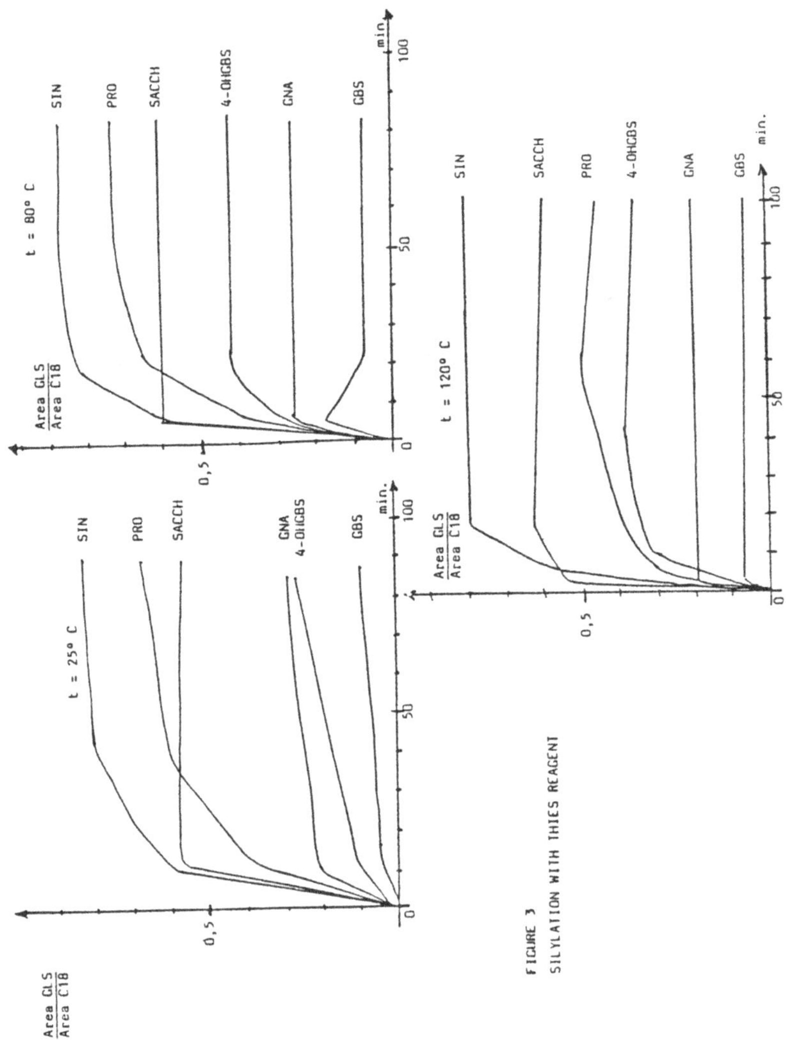

FIGURE 3
SILYLATION WITH THIES REAGENT

With THIES's reagent, the temperature has no meaningful effect ; only 4-hydroxylglucobrassicin presents a higher percentage of silylation at 80°C.

A complementary experiment was carried out taking into account the optimum periods of time which had been defined previously for each reagent, i.e. 10, and 40 min., with the three silylation-temperatures. All the samples were extracted simultaneously and silylations performed on aliquots.

Results (Table 1) confirm that with FENWICK's reagent, at 80°C, a maximum of silylated glucosinolates is generally obtained. However, the differences with other temperatures are not very important. We also note a smaller silylation at room temperature for 4-hydroxyglucobrassicin.

With THIES's reagent,the observed differences between two temperatures are quite low, except for 4-OH-GBS. Nevertheless, in this case, we note the presence of a non-identified peak (Fig.4) which could be partly silylated 4-OH-GBS. As a matter of fact, if we add both peaks, we obtain a value near to that measured at other temperatures.

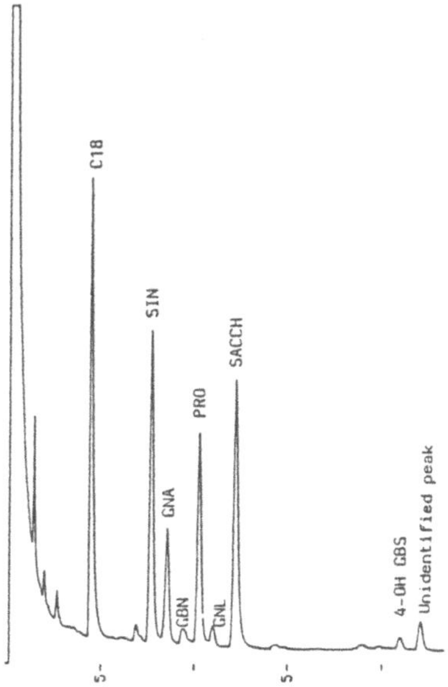

FIGURE 4

CONCLUSION

The previously described experiments show that gas chromatography with a temperature-programming may give us the possibility to determine glucosinolates in rapeseed seeds, indolglucosinolates included. However, the absence of pure controls did not allow us to identify small peaks which could correspond to other minor glucosinolates (neoglucobrassicin, gluconasturtin...). They are analysed by HPLC.

On the other hand, it was not possible to perform a quantitative comparison of results obtained by gas chromatography and of those obtained by HPLC,- the absence of pure controls making it impossible to know the coefficients of reaction for each technique. It would be necessary to solve this problem quickly to conclude validly.

In comparison with FENWICK's method, we note that the temperature of silylation may be reduced to 80°C, which can but contribute to the stability of indolglucosinolates.

To conclude, we may say that even if gas chromatography is not as efficient as HPLC to determine glucosinolates, in particular to determine minor products, however, it can be useful to quantify most important glucosinolates. Therefore, and for the above mentioned reasons (costs of investments and products, skilfulness of manipulators), it would be a pity not to keep this method in the future as a possibility for HPLC-unequipped laboratories.

REFERENCES

HEANEY R. K. and FENWICK G. R., 1980
The analysis of glucosinolates in Brassica species using gas chromatography. Direct determination of the thiocyanate ions precursors, Glucobrassicin and Neoglucobrassicin
J. Sc. Food Agri., 31, 593-599

HELBOE P., OLSEN O. and SORENSEN H., 1980
Separation of glucosinolates by high performance liquid chromatography
J. of Chromatography, 197, 199-205

MOLLER P., PLOGER A. and SORENSEN H., 1984
Quantitative analysis of total glucosinolates content in concentrated extracts from double low rapeseed by the Pd-glucosinolate complexe method.
Advances in the production and utilisation of cruciferous crops, vol. 11, p. 97-110, H. SORENSEN editor

QUINSAC A., RIBAILLIER D., 1984
Quantitative analysis of glucosinolates in rapeseed seeds optimization of desulphation.
Advances in the production and utilisation of cruciferous crops, vol. 11, 85-96, H. SORENSEN (editor). Martiners Nijhoff, Dr W. Junk Publishers

RUGRAFF L., CHEMIN-DOUAUD S. et KARLESKIND A., 1986
Dosage des glucosinolates dans les graines de crucifères. Application aux graines de colza et à leurs dérivés. Comparaison des méthodes en vue de leur automatisation.
Revue Française des Corps Gras, 33, 207-215

SPINKS E. A., SONES K. and FENWICK G.R. 1983
The quantitative analysis of glucosinolates in cruciferous vegetables, oilseeds and forage crops using high performance liquide chromatography.
Fette Seifen Anstrichmittel, n° 6, 228-231

THIES W., 1980
Analysis of glucosinolates via "on column" desulphation. Proceeding of symposium "analytical chemistry of rapeseed and its product". Winnipeg, 66-71

THIES W., 1983
Optimization of the "on column" desulphation and gas chromatography of glucosinolates.
Actes du 6ème Congrès International sur le Colza. Paris, 1343-1349

DETERMINATION OF GLUCOSINOLATES IN CRUCIFER VEGETABLES FOR HUMAN CONSUMPTION AND RAPESEED

B.G. MUUSE & H.J. VAN DER KAMP

RIKILT
State Institute for Quality Control of Agricultural Products
Bornsesteeg 45
6708 PD WAGENINGEN
The Netherlands

ABSTRACT

Glucosinolates (GSL) belong to non-nutritional natural compounds and occur in Crucifers. Rapeseed containing high amounts of GSL, 100 micromol per gram seed, is fed to cattle. The European Community stimulates the cultivation of double zero rapeseed with GSL content below 20 micromol/g seed.

In Crucifer vegetables GSL amounts of 10-40 micromol/g dry matter are found. The Indolyl GSL are found to activate detoxifying enzymes, resulting in an antimutagenic effect and may therefore be of nutritional interest.

The compatibility of analytical methods applying GCC, HPLC and fotometry is studied. The HPLC method using the desulfated glucosinolates is shown to be generally applicable and accurate when using respons factors.

The occurrence of GSL in vegetables and in a cultivar line of Lingot is tabulated.

INTRODUCTION

From literature a lot is already known concerning the determination and the occurrence of glucosinolates in nature. Reviews are recently made by Fenwick et al. (1983) and Mc.Gregor et al. (1983). Progess in analytical technics however asks for further investigations.

The general molecular structure of glucosinolates is formed by a beta thioglucose group, a sulfonated oxim midpart and an organic side chain R (fig. 1). The differentiation between glucosinolates is made by [illegible] side chain [illegible]

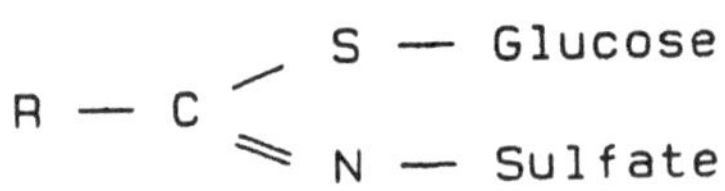

Figure 1. General molecular structure of GSL.

Some 100 glucosinolates have been described now. The major constituents can be classified as alkyls, aryls and indolyls. The alkyl types can also contain extra sulfur atoms which leads to sulfinyl, sulfonyl or just thio glucosinolates.

Glucosinolates are suggested to be chemotaxonomic compounds being growth regulating chemicals. With these glucosinolates crucifer varieties can be classified.

The biosynthesis of glucosinolates is possible from amino acids, with oxidative decarboxylation and sulfatisation steps and the sulfur in the cyanogenic midpart is thought to be introduced directly from cystein.

NUTRITIONAL EFFECTS OF GLUCOSINOLATES

Goitre

In animal feed the use of rapeseed is limited mainly by the effect on the thyroid gland (Fenwick & Heaney, 1983). Also other effects on adrenal gland, kidney and liver are known and in poultry, fishy taint and decolorations of the eggs have been noticed (Fenwick et al., 1981).

The thyroid enlarging effect by animals is caused by the very high content of glucosinolates in rapeseed. A content of 60 to 120 u Moles per gram seed is found which is equal to some 2% to 5%. This limits the use of rapeseed in animal feed and necessitates the cultivation of rapeseeds with low glucosinolate content.

The thyroid enlarging effect is mainly due to vinylthiooxazolidon which is a break down product of the hydroxybutenyl glucosinolate (known as progoitrin) and being the major component in rapeseed. This compound interfers with the thyroxine synthesis in the thyroid gland.

Antimutagenic effect

An interesting reason to study the glucosinolates is the antimutagenic propertie of some of them as is mentioned by Wattenberg (1978, 1979), Loub (1975), Pantuck (1976) and Hendrich (1983).

Recently Jongen and others (1986, in press) have developed and applied an IN VITRO method for the determination of modulating effects by indolyL GSL breakdown products. Protective effects of the indolyl compounds were found when testing the genotoxicity of known mutagenics like benzpyrene and dimethyl-nitrosamine.

Nematocide

Another reason to study the glucosinolates is the indication that glucosinolates could have a nematocidic action which might be of importance for the cultivation of sugar beets and potatoes (Forrest et al., 1983). This nematocidic action is thought to work by liberation of the glucosinolates from the roots thereby releasing hydrolysed glucosinolates into the ground. These liberated thiocyanates act like the nowadays used metamsodium.

Denaturation of skimmed milk powder

Finally, regulations from the European Community forced us to analyse the glucosinolates. To denaturate skimmed milk powder the Community regulated the addition of rapeseed meal with a high amount of glucosinolates, thus making the milk powder only usable for animal adults and not for young animals like baby pigs. The aim of this regulation is to reduce the skimmed milk powder stock without disturbing the existing market.

ANALYTICAL PROCEDURES

Hydrolysed glucosinolates

A lot of possibilities are known for analysing breakdown products. They can be obtained by hydrolysis with the enzyme myrosinase from white mustard seed. The oldest method is the argentometric titration of the distilled volatile isothiocyanates and is still the official but old fashioned directive of the European Community for the control on animal feed. Also for the determination via glucose reliable methods exist, leading to the total content of glucosinolates.
The most common method nowadays using myrosinase hydrolysis is given by the ISO 5504 method. This gaschromatographic method determines the liberated individual aglucones without any derivatisation. The disadvantange however of this method is that the hydroxy compounds mainly from progoitrin have to be determined separately with fotometry since they are not volatile enough for gaschromatography. Also the indolyl glucosinolates are not included in this method, which is a disadvantage especially for the "00" rapeseed.

Intact and desulfated glucosinolates

Since early studies have the disadvantage of using the hydrolysis step followed by uncontrolled autolyse reactions, there is a need for methods of analysis without hydrolysis in order to obtain better information on the toxic effects of GSL.

In literature several methods are published now which do not use the hydrolysis by myrosinase. Two principles are known, one using a clean up procedure and desulfatization of the GSL (Thies 1979; Spinks et al., 1984; Heaney & Fenwick 1980, 1982; Minchinton et al., 1982). The other proceeds without desulfatization and determines the intact glucosinolates (Thies 1976; Helboe et al., 1980).

When analysing the intact glucosinolates no reproducable results could be obtained by us either with HPLC or with the complexation method using palladiumchloride as chromophoor. On the contrary the desulfoglucosinolates which are obtained after enzymic desulfatization, could be determined very well with HPLC and the palladiumchloride method and to a lesser extent with gaschromatography. The palladiumchloride method measures only the total glucosinolate content whilst the chromatographic methods give the contents of individual glucosinolates including the hydroxyls and indolyls.

Principle of the method

A 1 ml hot water extract of the glucosinolates from a 100 mg dry and grounded sample is brought on a sephadex anion exchanger in a 1 ml Eppendorff pipettip and then washed.

The glucosinolates are cleaved with a desulfatase enzyme from the sulfate group which is fixed to the anion exanger and the desulfoglucosinolates are eluted from the column. This procedure is described in the official community directive for the determination of glucosinolates in rapeseed.

The desulfoglucosinolates can be used directly for HPLC analysis or can be used for gaschromatographic analysis after evaporation of the solvent followed by silylation of the hydroxy groups.

For the gaschromatographic separation we use a 25 m fused silica capillary column of CP SIL 5CB and for the HPLC separation a RP 18 reversed phase column. The elution of the desulfoglucosinolates from the reversed phase column is performed by a gradient system of acetonitril and water. An interesting advantage of the HPLC is the possibility to collect the compounds separately for further studye. Drying the fractions followed by separation on the HPLC system showed no alteration in the composition so proving that the desulfoglucosinolates can be obtained in dry form.

Comparison of methods

Comparison of the different methods lead to satisfying results when taken into account the expected differences of 100 to 400 percent (Figure 2).

For this comparison the GCC method was taken as a reference. The ISO 5504 method determines about 80% of the total GSL content since it does not measure the indolyls and therefore will be systematically lower than the other methods.

The results with the HPLC method compared with the GCC method initially differed widely, which was caused by the very great differences in molar extinction of the indolyl glucosinolates compared with other compounds.

Fig. 2 Comparison of methods related to the GCC method. HPLC values are obtained with and without respons factors

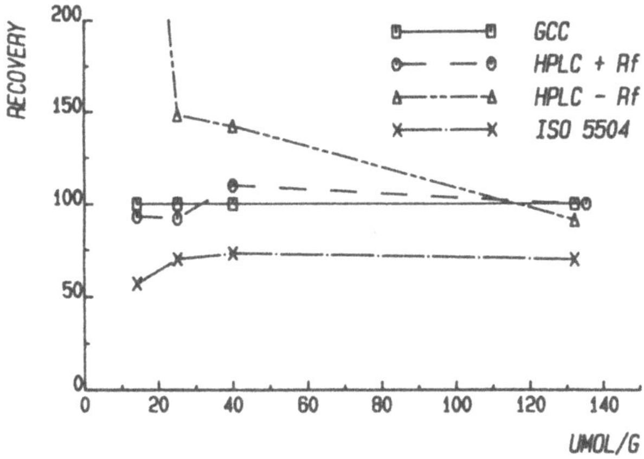

Respons factors for HPLC

The respons factors were obtained from the molar extinction coefficients and were measured by comparing the HPLC with the GCC results and by determining the collected HPLC fractions with palladium-chloride fotometry. This brought us to the conclusion that the different alkyl glucosinalates have the same extinction coefficient, but that the indolyl glucosinolates have an extinction coefficient some 4 to 5 times higher and depends from the wavelength (Figure 3).

Fig. 3 Molar extinction coefficients of some GSL types relative to sinigrin and related to the wavelength.

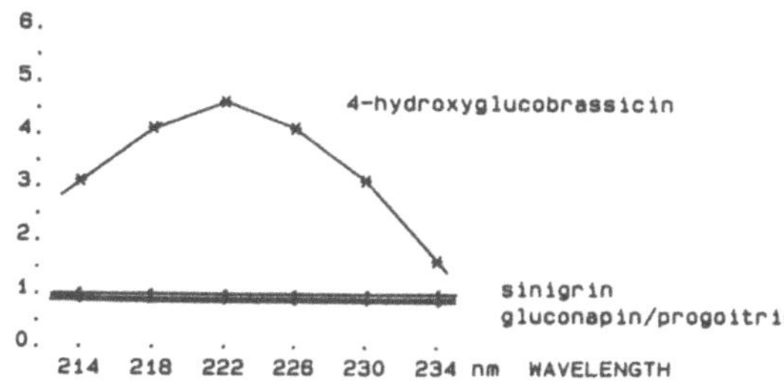

It was also found that the glucosinolates with an extra sulfur atom in the side chain have a two times higher extinction coefficient as the alkyls.

The absorbtion maximum for the indolyl glucosinolates is found at 222 nm (Figure 4). The indolyl and aryl glucosinolates have a second adsorption area after 260 nm, which gives a possibility for the classification of unknown peaks.

Fig. 4 UV-spectrum of some desulfo-GSL. Measured with diode array detection.

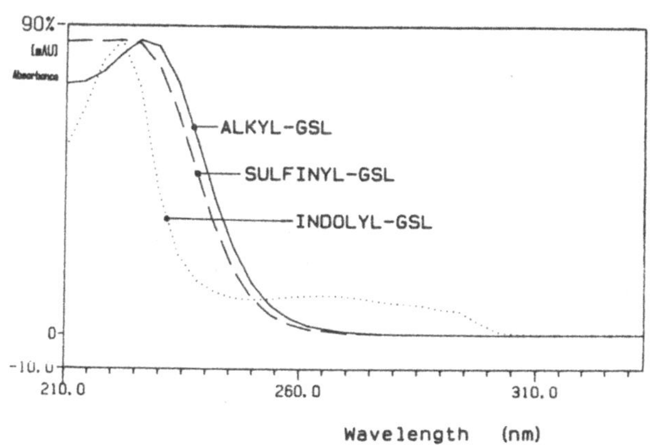

APPLICATION OF THE METHODS

"00" rapeseed

In the European Community an upper limit for the GSL content is fixed on 35 uMol per gram seed till 1988, and thereafter on 20 uMol per gram. This seems to be based on the development of double zero ("00") rapeseed in Canada as is published by Daun (1986). In Canada within two years the glucosinolate content in rapeseed was deminished from 100 to some 35 uMol per gram seed.

The official method in the European Community is the gaschromatographic method of the silylated desulfated GSL. Alternative methods leading to equivalent results may also be used.

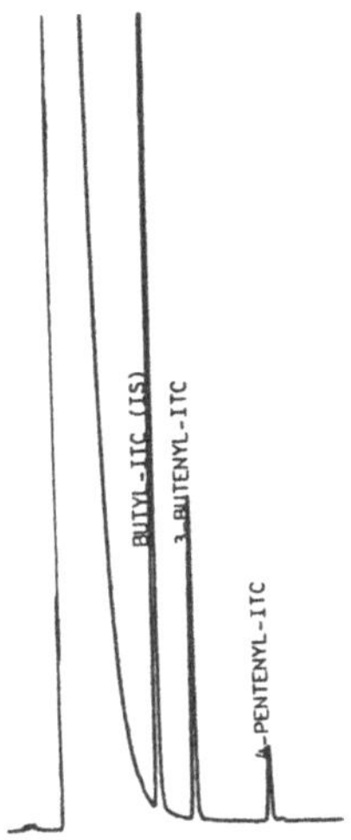

Fig. 5 Chromatogram of the ISO 5504 method.

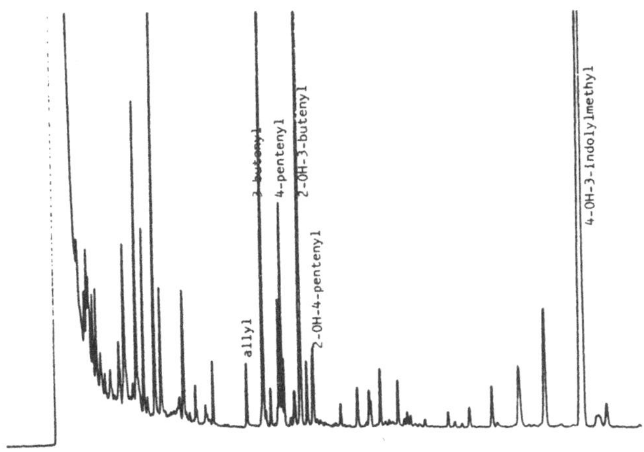

Fig. 6. Gaschromatogram of rapeseed GSL With the EEC method and using capillary column CP SIL 5CB 25 m.

With the ISO 5504 method which uses the myrosinase enzyme we found only butenyl and pentenyl isothiocyanate. They were lowered remarkably in "00" rapeseed (fig. 5). The official Community method gives a much more complex chromatogram partly caused by artefacts due to the silylation procedure. But there were also more glucosinolate compounds found with this method (fig. 6). For instance the hydroxy-butenyl and hydroxy-pentenyl GSL are found and at the end of the chromatogrammes the hydroxy-indolylmethyl glucosinolate appears.

The lowering of the different glucosinolates in "OO" rapeseed is proportional to the lower total GSL content. The indolyl glucosinolates however remain unchanged. In Table 1 the results of the glucosinolate composition in a cultivar line from Lingot is shown (Brzezinsky, Mendelewski, Muuse, 1986). Lingot is the normal glucosinolate variety whereas the BOH-484 variety is the ultimate "OO" rapeseed.

The results were obtained with HPLC after desulfatization of the glucosinolates. Tandem just belongs to the "OO" rapeseed definition till 1988. The hydroxy-indolylmethyl glucosinolate compound in the "OO" variety is increased relatively to the other compounds and goes upto 50% of the total GSL content.

The values are expressed on basis of dried meal. On basis of seed the values must be multiplied by 0.6.

Table 1. Content of glucosinolates in rapeseed varieties of a cultivar line from Lingot expressed in u Mol/g defatted meal. The total glucosinolate content is determined by HPLC of the desulfo-GSL and by thymol colorreaction on glucose.

Variety	THYMOL Total	Glucosinolate content HPLC Total	PRO	GNA	HGB	rest
Lingot	94.6	100.9	71.3	22.7	0.6	6.3
Tandem	59.3	59.2	40.4	13.0	3.3	2.5
Darmor	49.9	54.5	36.8	12.2	3.9	1.6
Linglandor	40.3	40.3	21.0	8.0	5.3	6.0
BOH 183*	25.1	24.5	12.6	4.9	5.1	1.9
Jantar	19.8	19.3	10.2	4.0	4.0	1.1
BOH 384	17.5	17.6	9.2	3.5	3.8	1.1
BOH 183*	13.3	15.5	6.9	2.7	4.7	1.2
BOH 484	9.7	11.4	4.0	1.5	5.1	0.8

* FROM DIFFERENT SOIL AND CLIMATE REGIONS
PRO=Progoitrin
GNA=Gluconapin
HGB=4-Hydroxy Glucobrassicin
(In Press: Cruciferea Newsletter 1986)

Crucifer vegetables

The interest to study glucosinolate in vegetables is founded by the goitre effect of mainly progoitrin and by the antimutagenic properties of the indolyl glucosinolates. In fig. 7 an HPLC chromatogram of a vegetable mixture gives an impression of the occurring glucosinolates. Only the minor constituents belong to the unknown peaks. The indolyl GSL seem to be present rather abundant but the indolyl GSL in the ultraviolet detector have a respons factor of about 4 at 226 nm which means that the peak areas had to be divided by 4 to obtain the results in mass percentages.

In general, the most occurring glucosinolates are progoitrin and glucobrassicin. In table 2 the results for some vegetables are given. The values are obtained from only a few samples and the natural variation in content is very high, so this result gives only a general idea of the glucosinolate contents in vegetables.

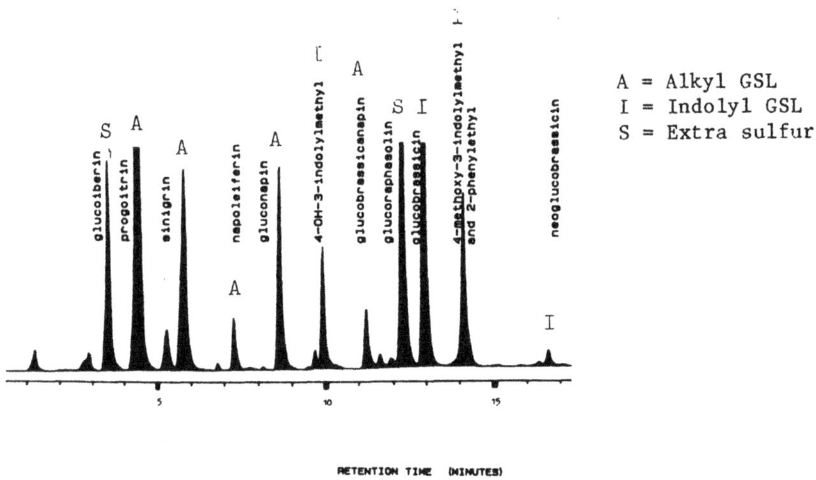

Fig. 7 HPLC chromatogram of desulfo-GSL in a mixture of vegetables

When the vegetables were cooked, we found that the content of glucosinolates is decreased with some 25% without a change of the composition, which means that all glucosinolates were lost to the same extent. The way of cooking, normally or with pressure, did not influence the results.

The products in table 2 are arranged in decreasing order of the total glucosinolate content and vary from 42 to 11 uMol per gram dry food which corresponds to 1.5 to 0.5 percent. Common vegetables with a high total glucosinolate content and also high indolyl glucosinolates are Savoye cabbage, Cauliflower and Brussels sprouts.

Table 2. Glucosinolate content of Crucifer vegetables expressed as u Mol/g dry matter and effect of cooking by determining the recovery percentage of the remaining GSL in the food.

CRUCIFER:	uMOL/g dry matter Total	Alkyl	Indolyl	+Sulfur	Rest	Cooking recovery %
Savoy cabbage	42	13	13	13	3	65
Rettich	29	0	2	25	2	-
Brussels sprouts	25	13	6	3	3	70
Cauliflower	23	4	10	1	8	75
Red cabbage	20	14	2	4	0	70
Curley kale	19	7	3	7	2	45
White cabbage	16	8	3	5	0	70
Chinese kale	13	5	6	1	1	80
Turnip	11	5	4	0	2	-

A typical product containing glucosinolates with extra sulfur is Rettich. The pungent taste of this product is attributed to the breakdown products of these glucosinolates.

CONCLUSIONS

The HPLC method for the analysis of desulfo-glucosinolates, is generally applicable and offers the possibility of collecting the glucosinolate fractions for further study. The respons of the indolyl glucosinolates in the ultraviolet detector is some 4 to 5 times higher than the alkyls and also the glucosinolates with extra sulfur in the molecule have a higher respons factor.

Cultivation of rapeseed may decrease the alkyl GSL content, which enables a higher consumption of rapeseed. The indolyl GSL content of rapeseed remain constant and increases up to 50% of the total GSL in "00" rapeseed.

IN VITRO investigations by the Netherlands Agricultural University demonstrate antimutagenic effects of indolyl glucosinolate breakdown products.

Savoy cabbage and cauliflower contain high amounts of the indolyl glucosinolates, about 0.5% on dry matter. Rettich contains more than 1% of sulfur glucosinolates.

Some 25% of the glucosinolates are lost by normal or pressure cooking without changing the composition of the remaining glucosinolates in the vegetable.

REFERENCES

Brzezinski, W., Mendelewski, P., Muuse, B.G. 1986. Comparative study on determination of glucosinolates in rapeseed. Cruciferae Newsletter, in press.

Daun, J.K. 1986. Glucosinolate levels in western canadian rapeseed and canola. J.Am.Oil Chem. Soc, 63, 639-643.

Fenwick, G.R., Heaney, W.J. and Mullin, W.J. 1983. Glucosinolates and their breakdown products in food and food plants. Critical Rev. in Fd. Sc. & Nutr., 18, 123-201.

Fenwick, G.R., Heaney, W.J. 1983. Glucosinolates and their breakdown products in cruciferous crops, foods and feedingstuffs. Fd. Chem., 11, 249-271.

Fenwick, G.R., Pearson, A.W., Greenwood, N.M., Butler, E.J. 1981. Rapeseed meal tannings and egg taint. Animal Feed Sc. & Techn., 6, 421-431.

Forrest, J.M.S., Farrer, L.A. 1983. The respons of eggs of the white potato cyst nematode globdera pallida to diffusate from potato and musterd roots. Ann. Appl. Biol., 103, 283-389.

Gregor, Mc D.I., Mullin, W.J., Fenwick, G.R. 1983. J. Ass. Off. Anal. Chem., 66, 826-849.

Heaney, R.K., Fenwick, G.R. 1980. The analysis of glucosinolates in Brassica species using gaschromatography. J. Sci. Fd. Agric., 31, 593-599.

Heaney, R.K., Fenwick, G.R. 1982. The quantitative analysis of indole glucosinolates by gaschromatography. J. Sci. Fd. Agric., 33, 68-70.

Helboe, P., Olsen, O., Sorensen, H. 1980. Separation of glucosinolates by HPLC. J. of Chrom., 197, 199-205.

Hendrich, S., Bjeldanes, L.F. 1983. Effects of dietary cabbage, Brussels sprouts, Illicum verum, Schizandra chinensis and alfalfa on the benzo(a)pyrene metabolic system in mouse liver. Fd. Chem. Toxic., 21, 479.

Jongen, W.M.F., Topp, R.J., Tiedink, H.G.M. and Brink, E.J. 1986. A co-cultivation system as a model for IN VITRO studies of modulating effects of naturally occurring indoles on the genotoxicity of model compounds. Fd. Chem. Toxicol. In press.

Loub, W.D., Wattenberg, L.W. and Davis, D.W. 1975. Aryl hydrocarbon hydroxylase induction in rat tissues by naturally occurring indoles of cruciferous plants. J. Natl. Cancer Inst., 54, 985.

Minchinton, I. Sang, J., Burke, D. Truscott, J.W. 1982. Separation of desulfoglucosinolates by reversed phase HPLC. J. of Chrom., 247, 141-148.

Pantuck, E.J., Hsiao, K.C., Loub, W.D., Wattenberg, L.W., Kuntzman, R., and Conney, A.H. 1976. Stimulatory effect of vegetables on the intestinal drug metabolism in the rat. J. Pharm. Exp. Ther., 1978, 277.

Spinks, E.A., Sones, K., Fenwick, G.R. 1984. The quantitative analysis of glucosinolates in cruciferous vegetables, oilseeds and forage crops using HPLC. Fette Seifen Anstrichmittel, 6, 228-231.

Thies, W. 1976. Quantitative gaschromatography of glucosinolates on a microliter scale. Fette Seifen Anstrichmittel, 6, 231-234.

Thies, W. 1979. Detection and utilization of a glucosinolate sulfohydrolase in the edible snail, helix pomatia. Naturwiss., 66, 364-365.

Wattenberg, L.W. 1979. Inhibition of carcinogenic effects of polycyclic hydrocarbons by benzyl isothiocyanate and related compounds. J. Natl. Cancer Inst. 58, 395.

Wattenberg, L.W., Loub, W.D. 1978. Inhibition of polycyclic aromatic hydrocarbon-induced neoplasia by natural occurring indoles. Cancer Res., 38, 1410.

APPROACH TO DETERMINATION OF HPLC RESPONSE FACTORS FOR GLUCOSINOLATES

R. Buchner

Institute for Agronomy and Plant Breeding
University of Göttingen
von Siebold Strasse 8
D-3400 Göttingen
Federal Republik of Germany

ABSTRACT

A procedure involving the "Thymol-test" proposed by McGregor has been used for the indirect determination of HPLC response factors at 229 nm. Problems occuring by using this indirect method for calibration are discussed. Taking glucosinolates' spectrum and the different approaches into account a comparision of all till now published HPLC response factors is made. The influence of different solvent systems and different HPLC equipment is considered and its importance for the establishment of internationally valid factors is discussed.

INTRODUCTION

Brzezinski et al. (1984) established the "Thymol-test" for glucosinolates' analysis. Because of the lack of purified glucosinolates McGregor (1985) proposed the use of the "Thymol-test" for calibration of the HPLC of glucosinolates. McGregor received response factors quite similar to those determined at the same wavelength by Sang and Truscott (1984) using purified glucosinolates.

A handicap for comparison of all the response factors determined worldwide till now is the use of different detection wavelengths and the poor knowledge about way and quantity of influence through HPLC parameters like gradients' shape, acetonitril content, temperature, HPLC equipment a.s.o..

Furtheron it is still unknown, whether it should be preferred to establish internationally valid HPLC response factors for glucosinolates' analysis or to enable every laboratory to calibrate its HPLC on its own. In both cases it is still a unsolved problem, how to do it.

This paper reports problems and results of HPLC calibration using the "Thymol-test", tries to quantify different parameters of influence on relative response factors and to step forward in international HPLC calibration of glucosinolates.

MATERIALS AND METHODS

(a) Equipment for extraction and purification of glucosinolates see Buchner (this volume).

(b) UV spectrophotometer.

(c) HPLC chromatograph: dual pump gradient instrument with a UV detector (two Waters M510 pumps, Waters WISP 710b, Data Module M730, System Controller M721 and UV detector M441 with the atomic emission light of a cadmium lamp (229 nm) as source).

(d) LC colum - A C18 reversed phase column Waters Novapak 150 x 3.9 mm, 4 um particle size.

Reagents

(a) Reagents for extraction and purification of glucosinolates see Buchner (this volume).

(b) glucose solution 0.5 mmol/l.

(c) 1 % (w/v) thymol in ethanol.

(d) 77 % (v/v) sulfuric acid

(e) ultra pure water and acetonitrile for HPLC

Calibration of HPLC for glucosinolates using the "Thymol-test"

(1) Freeze drying of circa 10 pooled samples of desulfo glucosinolates prepared as described by Buchner (this volume).

(2) Injection of an aliquot into HPLC. Used program: 0-16.5 min 0 - 16.5 % acetonitrile, 16.5 - 18 min 16.5 - 0 % acetonitrile (rest:water); flow: 2.0 ml/min.

(3) Collection of the single peaks behind the detector.

(4) Freeze drying of the samples.

(5) Resolving in 1 ml water.

(6) Adding 7 ml of the sulfuric acid and 1 ml of the thymol-solution.

(7) Mixing vigorously.

(8) 35 minutes at 100°C

(9) Cooling slowly the now uncapped tubes.

(10) Absorption at 505 nm against a blank (pure water instead of glucosinolate sample).

(11) Calibration of thymol reaction using different amounts of the glucose solution (filled up to 1 ml with water).

(12) Calculation of absolute response factors:

$$\frac{\text{umol glucose}}{\text{integrated area}} = \text{absolute response factor}$$

(13) Calculation of relative response factors setting sinigrin to 1.0.

RESULTS

Determined response factors

TABLE 1 HPLC response factors for some glucosinolates at 229 nm determined indirectly by use of the "Thymol-test". The absolute response-factors refer to conversion in umol glucosinolates for the HPLC machine in Göttingen.

glucosinolate	number of determ.	s%	absolute R_F-value	relative R_F
Iberin	3	7.6	1.25518×10^{-9}	1.07
Progoitrin	13	4.1	1.27933×10^{-9}	1.09
Sinigrin	16	4.7	1.17242×10^{-9}	1.00
Gluconapin	19	5.9	1.30683×10^{-9}	1.11
4OH-Glucobrassicin	10	10.2	0.33120×10^{-9}	0.28
Glucobrassicanapin	7	5.5	1.34634×10^{-9}	1.15
Glucotropaeolin	13	5.1	1.11111×10^{-9}	0.95
Glucobrassicin	11	6.8	0.33988×10^{-9}	0.29
Neoglucobrassicin	5	5.5	0.23869×10^{-9}	0.20

The increased standard deviation of 4-hydroxy-glucobrassicin probably is caused by the strongly temperature dependent detection of this compound in the HPLC (see also Buchner, this volume).

<u>Problems of this approach</u>

Besides the problems caused by the high susceptibility of the "Thymol-test" to interference, e.g. of detergents, the following important problems were observed:

o After the freeze drying step some glucosinolates were found partly at the cap of the loosely capped tubes. It is assumed, that perhaps a small quantity could escape from the tubes. This problem was especially found for neoglucobrassicin. Possibly this "evaporation effect" affects different glucosinolates to different extent. May be the same problem occurs when evaporation is used to take samples to dryness as McGregor (1985) did.

o It's only proved for sinigrin, that there isn't any remarkable influence of the aglucon part of the glucosinolate on the thymol reaction (Brzezinski et al., 1984). For all the other glucosinolates the influence isn't known.

At the moment experiments are done comparing results from the use of the "Thymol-test" and from the "Hexokinase glucosetest" for analysis of glucosinolates. There will be some research on the escape of glucosinolates during freeze drying or evaporation of the solvent. Furthermore for comparison of the different approaches the response factors for progoitrin, sinigrin, glucotropaeolin, glucobarbarin and glucobrassicin will be determined using these compounds purified.

COMPARISON OF ALL TILL NOW PUBLISHED RELATIVE RESPONSE FACTORS FOR HPLC OF GLUCOSINOLATES

TABLE 2 HPLC relative response factors for glucosinolates. The factors mean desulfo glucosinolates except the last column (intact glucosinolates).

	Truscott (1984) 226 nm	McGregor (1985) 226 nm	Nyman (1986) 228 nm	Buchner 229 nm	Spinks (1984) 230 nm	Möller (1985) 235 nm
Capp.						1.16
Iberin		1.06		1.07	1.50	1.09
Raphanin						1.12
PRO	1.01	1.01 (S) 0.96 (R)	1.00 (former 0.91)	1.09	1.88	1.04
SIN	1.00	1.00	1.00	1.00	1.00	1.00
GNL	1.02		1.00			1.06
Alyss.		0.99				1.12
Sinalb.	0.51	0.50				0.52
GNA	1.08	0.94	0.96	1.11	1.63	1.09
4OHGBC	0.27	0.24	0.54	0.28		0.27
Ib.vir.		1.18				
GBN	1.08	0.99	1.00	1.15		1.07
GTL	0.92	0.86		0.95	1.25	0.88
ERU	1.01					
GBC	0.29	0.26	0.35	0.29	0.75	0.45
Nast.		0.95				
Barbarin						0.95
4MEGBC		0.25				0.27
NEOGBC		0.23		0.20		0.37

Capp.=glucocapparin; PRO=progoitrin; SIN=sinigrin; GNL=gluconapoleiferin; Allyss.= Glucoallyssin; GNA=gluconapin; 4OHGBC=4-hydroxy-glucobrassicin; Ib.vir.=glucoibervirin; GBN=glucobrassicin; Nast.=gluconasturtiin; 4MEGBC=4-methoxy-glucobrassicin; NEOGBC=neoglucobrassicin

A comparison with the response factors of the intact glucosinolates' analysis (last column) is not possible because of the totally different environment in the HPLC (methanol instead of acetonitrile, much higher concentrations of the stronger eluent, presence of a counter ion) and the different molecular structure of the glucosinolates.

The use of different wave lengths is not a problem for the alkenyl glucosinolates but it is one especially for the indolyl glucosinolates with their steep slope of absorbance between 230 and 225 nm. It should be mentioned , that the use of normal filter spectrophotometer because of their little precise wave length adjustment is problematic.

Besides these handicaps and the influence of different HPLC equipments it is evident now, that the response factors relative to sinigrin are about 1 for the alkenyl and the sulfur glucosinolates, about 0.9 for benzyl- and phenylglucosinolates and about 0.25 for indolylglucosinolates.

To compare values obtained at the wave lengths 226 and 229 nm the absorption spectrum of some glucosinolates was taken and the factors for conversion of response factors from one wave length to the other for each glucosinolate were calculated. Figure 1 shows typical absorption lines of the different groups of glucosinolates.

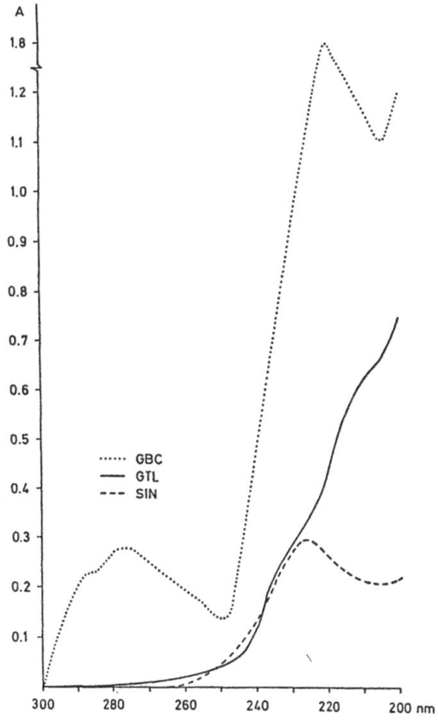

Fig. 1 Absorption spectrum of identical amounts of glucobrassicin, glucotropaeolin and sinigrin

The spectra of other representatives of the respective glucosinolate group show spectra, which are very similar in their shape but different in details like the exact position of the absorption maximum.

TABLE 3 Detail of table 2 with the response factors obtained at 229 nm converted by calculation based on the spectra taken (fig. 1) to 226 nm.

	Truscott (1984) 226 nm	McGregor (1985) 226 nm	Buchner calculated 226 nm	Buchner 229 nm
iberin		1.06	1.00	1.07
progoitrin	1.01	1.01 (S) 0.96 (R)	1.08	1.09
sinigrin	1.00	1.00	1.00	1.00
gluconapin	1.08	0.94	1.07	1.00
4-hydroxy-glucobrassicin	0.27	0.24	0.20	0.28
glucobrassicanapin	1.08	0.99	1.13	1.15
glucotropaeolin	0.92	0.86	0.90	0.95
glucobrassicin	0.29	0.26	0.20	0.29
neoglucobrassicin		0.23	0.17	0.20

Datas of alkenyl glucosinolates and glucotropaeolin look acceptable, but the values for the indolyl glucosinolates don't satisfy. Perhaps except neoglucobrassicin (see above) the reason is unknown.

PARAMETERS OF POSSIBLE INFLUENCE ON RELATIVE RESPONSE FACTORS

For desulfo glucosinolates there wasn't any influence of the below listed parameters found:

- o acetonitrile content of the mobile phase
- o temperature of the column, except 4-OH-glucobrassicin (see Buchner, this volume)
- o acetonitrile/water gradient slope
- o different flow speeds can be eliminated by simple calculation:

area (2) = area (1) x speed (2) / speed (1)
The following parameters may have some different influence to different glucosinolates:

o LC column material; may influence especially the very early and the late eluted compounds.

o HPLC equipment; e.g. when pumps are used for extremely low speeds they are not designed for. Mainly the very early eluting compounds (iberin, progoitrin, sinigrin) might be influenced.

There is further need for research on these two parameters.

CONCLUSIONS

There are two paths to go to enable people to calibrate their HPLC for analysis of glucosinolates. First to determine response factors for each HPLC used for analysis of glucosinolates, second to fix the analysis procedure as far as possible and to define relative response factors in an international agreement: - e.g. for alkenyl and sulfur glucosinolates 1, for indolyl glucosinolates 0.25 and for benzyl/phenyl glucosinolates 0.9 or an individual factor for each glucosinolate.

The first path can be used by applying pure glucosinolates or by providing standard rape seed samples, whose content of different glucosinolates is defined by international ring tests. As there are no pure glucosinolates available in an acceptable ammount, only the second possibility is left, which is supported by the author as the probably most exact calibration procedure.

It is still to be discussed, which response factors are used for the international ring tests.

Notwithstanding these problems we suggest to fix the detection wave length to 229 nm and to use whenever possible an atomic emission light as source. As internal standard the author supports the use of glucobarbarin and - if necessary - glucotropaeolin as substitute. By no means an early eluting glucosinolate like sinigrin or a non-glucosinolate should be used.

ACKNOWLEDGEMENTS

The author thanks Prof. G. Röbbelen and Prof. W. Thies for their support and helpful discussions during this work. He also thanks Miss K. Fischer and Mr. U. Ammermann for excellent technical assistance and in particular for drawing the figures. Financial support of the EC Commission is greatfully acknowledged.

REFERENCES

Brzezinski, W., Mendelewski, P. 1984. Determination of total glucosinolate content in rapeseed meal with thymol reagent. Z. Pflanzenzüchtg., 93, 177-183.

Buchner, R., 1986. Comparison of procedures for optimum extraction, purification and analysis of indolylglucosinolates. This volume.

McGregor, D.I., 1985. Determination of glucosinolates in Brassica seed. Eucarpia Cruciferae Newsletter, 10, 132-136.

Møller, P., Olsen O., Plöger, A., Rasmussen, K.W., Sørensen, H. 1985. Quantitative analysis of individual glucosinolates in double low oilseed rape by hplc of intact glucosinolates. In "Advances in the production and utilization of cruciferous crops with special emphasis to oilseed rape" (Ed. H. Sørensen), Copenhagen. p. 111-126.

Nyman, U. 1986. Personal communication.

Sang, J.P., Truscott, R.J.W. 1984. Liquid chromatographic determination of glucosinolates in rapeseed as desulfoglucosinolates. J. Assoc. Off. Anal. Chem., 67, 829-833.

Spinks, E.A., Sones, K., Fenwick, G.R. 1984. The quantitative analysis of glucosinolates in cruciferous vegetables, oilseeds and forage crops using high performance liquid chromatography. Fette, Seifen, Anstrichmittel, 86, 228-230.

ISOLATION OF INTACT GLUCOSINOLATES BY COLUMN CHROMATOGRAPHY AND DETERMINATION OF THEIR PURITY

Birthe Bjerg and Hilmer Sørensen

Chemistry Department,

Royal Veterinary and Agricultural University,

40, Thorvaldsensvej, DK-1871 Frederiksberg C, Denmark.

ABSTRACT

Isolation of intact glucosinolates requires knowledge of their occurrence, structures and properties.Of equal importance is knowledge of instrumentation, structures and properties of column materials required as well as knowledge of occurrence, structures and properties of co-occurring natural products, including the myrosinases (Thioglucoside glucohydrolase, EC 3.2. 3.1).

The purposes of glucosinolate isolation can be quite different. This can be isolation of small amounts on mini-columns for quantitative methods of analysis, isolation of pure reference compounds, isolation and identification of new compounds, and finally, isolation of appreciable amounts for feeding trials or other special purposes. Thereby, differences between recommendable techniques exist and need to be considered.

It is important to realize that it is nearly impossible to obtain 100% pure glucosinolates. For several reasons, including possibilities for performance of reliable quantitative methods of analysis, the exact and actual purity of the glucosinolates must be determined. Therefore, we need to consider the limitations and possibilities of different methods suitable for determination of purity of isolated glucosinolates. The data obtained in our laboratory show that it is recommendable to use several, of each other independent, methods.

This paper comprises a brief discussion and presentation of recent advances within the above mentioned points of interest in relation to glucosinolate isolation and methods required for determination of their purity.

INTRODUCTION

Glucosinolates co-occur with myrosinase isoenzymes (thioglucoside glucohydrolase EC. 3.2.3.1) (Buchwaldt et al., 1986) in all of the hitherto investigated plants collected in the order Capparales (Dahlgren, 1974; Rodman, 1981). A restricted number of other plant families have been shown to contain glucosinolates, mostly on the basis of identified degradation products of glucosinolates, but some of these reports have been seriously questioned (Larsen et al., 1983; Bjerg et al., 1986a).

Capparales comprises in the botanical broad sense: LIMNANTHACEAE; TROPAEOLACEAE; BRETSCHNEIDERACEAE*; SALVADORACEAE*; TOVARIACEAE; PENTADIPLANDRACEAE*; GYROSTEMONACEAE*; BATACEAE*; MORINGACEAE; BRASSICACEAE; CAPPARACEAE; RESEDACEAE.

The family Brassicaceae (Cruciferae) includes a number of crops important for human and animal nutrition, especially oilseed rape (Sørensen, 1985a). Plants belonging to the other families are, however, also important to consider in relation to the subject for the present study, as the first requirement is a suitable source of the glucosinolate in question.

The pungency, flavor and many undesirable toxic manifestations of different crucifer materials are associated with glucosinolates, affecting, e.g., the technically and economically important oils and proteins from these plants (Olsen and Sørensen, 1981; Eggum et al., 1985a). These properties of both glucosinolates and degradation products thereof are responsible for much of the interest in glucosinolates. Figure 1 shows structures and names of such compounds as are known to occur in varying amounts in cruciferous plant materials and their chemical, biochemical and physiological properties suggest that they may have especial significance _in vivo_.

The structural types of glucosinolates known and described are of course only those which can be isolated and identified by the available methods. Recent years research within this field has clearly shown a need for reconsideration of these problems. The cinnamoylderivatives of glucosinolates (Fig. 1) have been more or less neglected up to now owing to a lack in applied methods, even as they are quantitatively dominating among the glucosinolates in some plants or plant parts. This is also the case for other relatively unstable glucosinolates and a serious problem exists with most of the applied methods with respect to the possibilities of including glucosinolates with an acidic group in the side chain.

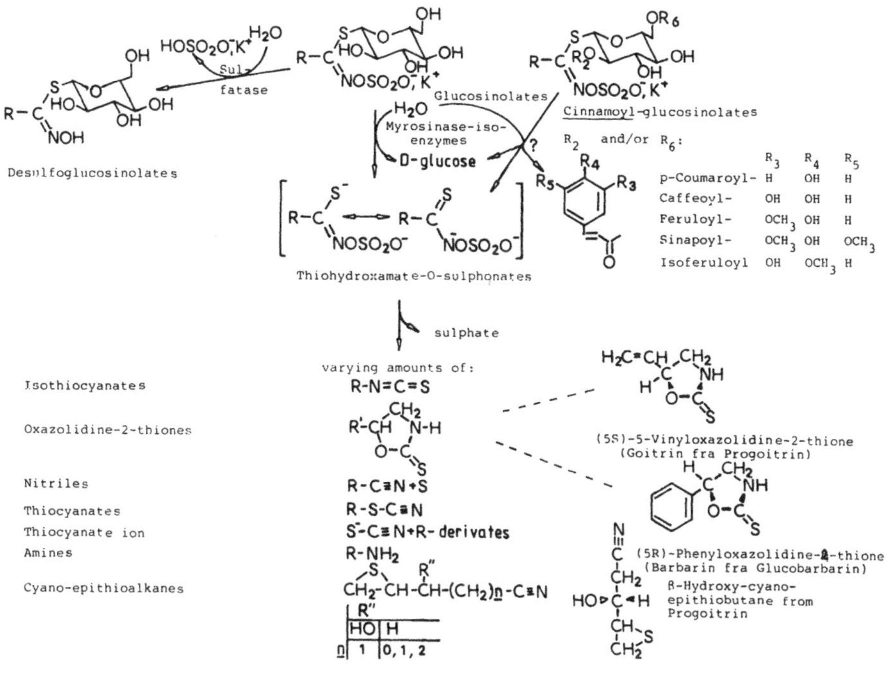

Fig. 1. Names and structures of degradation products of glucosinolates formed by sulfatase- and myrosinase-catalysed hydrolysis: R = side chain of glucosinolates; - structures and names of some selected glucosinolates are shown in Table 1. R_2 and/or R_6 are cinnamoyl-derivatives.

TABLE 1. Selected glucosinolates representing structures which need to be considered in discussions of quality of oilseed rape and different cruciferous materials and thereby requirements to methods of glucosinolate analysis.

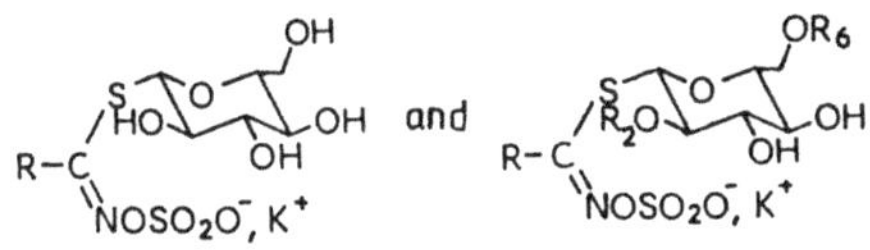

R_2 & R_6 = cinnamoyl-derivatives (Fig. 1)

Glucosinolates

No.	Structure of R group	Semisystematic names	Trivial names
1	CH_2=CH-CH_2-	Allylglucosinolate	Sinigrin
2	CH_2=CH-CH_2-CH_2-	But-3-enylglucosinolate	Gluconapin
3	CH_2=CH-CH_2-CH_2-CH_2-	Pent-4-enylglucosinolate	Glucobrassicanapin
4	CH_2=CH-CH(OH)-CH_2-	(2R)-2-Hydroxybut-3-enylglucosinolate	Progoitrin
5	-- " --	(2S)-2-Hydroxybut-3-enylglucosinolate	Epiprogoitrin
6	CH_2=CH-CH_2-CH(OH)-CH_2-	(2R)-2-Hydroxypent-4-enylglucosinolate	Napoleiferin
7	CH_3-S-CH_2-CH_2-CH_2-	3-Methylthiopropylglucosinolate	Glucoibervirin
8	CH_3-S-CH_2-CH_2-CH_2-CH_2-	4-Methylthiobutylglucosinolate	Glucoerucin
9	CH_3-S-CH_2-CH_2-CH_2-CH_2-CH_2-	5-Methylthiopentylglucosinolate	Glucoberteroin
10	CH_3-SO-CH_2-CH_2-CH_2-	3-Methylsulphinylpropylglucosinolate	Glucoiberin
11	CH_3-SO-CH_2-CH_2-CH_2-CH_2-	4-Methylsulphinylbutylglucosinolate	Glucoraphanin
12	CH_3-SO-CH_2-CH_2-CH_2-CH_2-CH_2-	5-Methylsulphinylpentylglucosinolate	Glucoalyssin
13*	CH_3-SO-CH=CH-CH_2-CH_2-	4-Methylsulphinylbut-3-enylglucosinolate	Glucoraphenin
14	CH_3-SO_2-CH_2-CH_2-CH_2-	3-Methylsulphonylpropylglucosinolate	Glucocheirolin
15	CH_3-SO_2-CH_2-CH_2-CH_2-CH_2-	4-Methylsulphonylbutylglucosinolate	Glucoerysolin
16	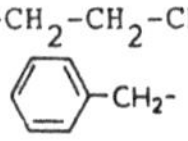	Benzylglucosinolate	Glucotropaeolin
17	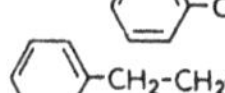	Phenethylglucosinolate	Gluconasturtiin
18*	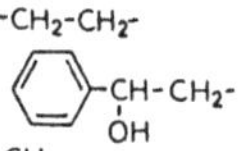	2-Hydroxy-2-phenylethylglucosinolate	Glucobarbarin
19	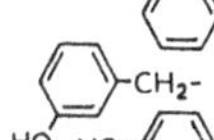	m-Hydroxybenzylglucosinolate	Glucolepigramin
20	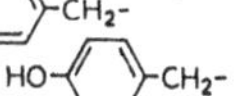	p-Hydroxybenzylglucosinolate	Sinalbin
21	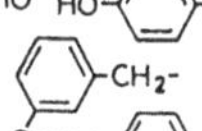	m-Methoxybenzylglucosinolate	Glucolimnanthin
22	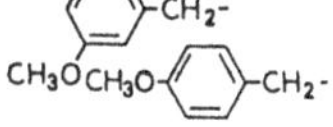	p-Methoxybenzylglucosinolate	Glucoaubrietin
	Indol-3-ylmethylglucosinolates:		
23	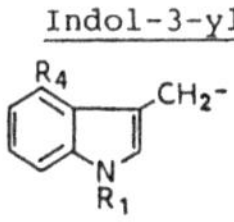 (R_1=R_4=H	Indol-3-ylmethylglucosinolate	Glucobrassicin
24	R_1=OCH_3; R_4=H	N-Methoxyindol-3-ylmethylglucosinolate	Neoglucobrassicin
25	R_1=SO_3^-; R_4=H	N-Sulphoindol-3-ylmethylglucosinolate	Sulphoglucobrassicin
26	R_1=H; R_4=OH	4-Hydroxyindol-3-ylmethylglucosinolate	4-Hydroxyglucobrassicin
27	R_1=H; R_4=OCH_3	4-Methoxyindol-3-ylmethylglucosinolate	4-Methoxyglucobrassicin

The glucosinolates No. 1-15 are biosynthetically derived from methionine. Those derived from phenylalanine (No. 16-22) are of interest as intern references.
*Occur also as cinnamoylderivatives (Fig. 1).

The actual or potential antinutritional or toxic properties of glucosinolates and degradation products thereof have been the subject of comprehensive studies based on isolated intact glucosinolates and myrosinases (Bille et al.,1983a and 1983b). The investigations comprise up to now ten different glucosinolates (Bjerg et al. 1986b; Eggum et al., 1985b). In addition to energy- and N-balance trials with rats, trials with pigs (Eggum et al., 1985c), young bulls (Andersen and Sørensen, 1985), mink (Sørensen, 1985b), insects (Larsen et al., 1985) and fungal pathogens (Buchwaldt et al., 1985) have also been included in these investigations. Studies of these problems, the metabolism of glucosinolates in plants (Larsen et al., 1984) and animals, chemical, biochemical and physiological properties of glucosinolates and their breakdown products as well as development of reliable methods for their analysis in foods (Olsen and Sørensen, 1981; Bjerg and Sørensen, 1986) all require purified glucosinolates often in relatively large amounts.

In the course of ongoing studies on glucosinolate problems (vide supra) it was necessary to obtain reasonably large quantities of certain glucosinolates. These compounds were not available, and synthesis of different glucosinolates according to known methods are of restricted value in relation to preparation of pure glucosinolates. Therefore, methods for the isolation of intact glucosinolates from plant material, preferably seeds,have been developed (Olsen and Sørensen, 1979). The separation and purification techniqies have been improved (Bjerg et al., 1984) especially by further improvements of the column chromatographic techniques as described in this paper which also includes descriptions and discussions of methods used for determination of glucosinolate purity.

MATERIALS AND METHODS

General methods and instrumentations used in the various steps have been described in detail previously (Olsen and Sørensen, 1979; Olsen and Sørensen, 1980; Bille et al., 1983b; Bjerg et al., 1984; Bjerg and Sørensen, 1986).

The plant material used for isolation of glucosinolates depends on the purpose of the study. For isolation of reference glucosinolates in appreciable amounts it is recommendable to select seed materials containing only the required glucosinolate (Møller et al., 1985a) or, failing that, a source containing as few as possible and then with maximum differences in their chemical properties (Olsen and Sørensen, 1981).

Crude glucosinolate extracts

The first steps in the extraction procedures differ somewhat according to the purpose. This depends on whether the source is appreciable amounts of green or vegetative plant materials (Olsen and Sørensen, 1979) or seeds (Bille et al., 1983b). It can also be semimicro methods (Bjerg et al., 1984) or quantitative methods of analysis (Bjerg and Sørensen, 1986).

To ensure inactivation of the myrosinases, it is important that the homogenizations are performed in a boiling extraction solution. Extraction in boiling methanol-water (7:3; three times) using an Ultra Turrax homogenizer gives efficient glucosinolate extraction and denaturation of macromolecules. The homogenates have to be centrifugated (3000 x g; 10 min; 30°C), and the supernatants need normally recentrifugation/filtration and concentration to a small volume (not dryness) to give a crude extract appropriate for column chromatographic isolation and purification of intact glucosinolates.

Group separations by ion-exchange column chromatography

The principles, column materials and ideas behind the group separation technique are described elsewhere (Bjerg et al., 1984). Following extraction, the glucosinolates are isolated by such technique (Figure 2) with modifications as described in the articles mentioned in the section general methods and instrumentation (vide supra).

This group separation technique is based on ion-exchange columns. The methods are, however, not a traditional ion-exchange technique. For the columns (A) and (C), the retained ions are released and eluted by use of an eluent (volatile) which removes the charge on the column materials. For the (B) column, it is the positive net charge on the retained compounds which are removed by the eluent (volatile), and use of the pre-column (A) is important. There are only ions on the (B) column which can be eluted by the gentle conditions. Even as it is very gentle separation principles, there are some compounds which can only be isolated by a modified technique (vide infra), and for simple quantitative methods of analysis, special adaptations are normally required.

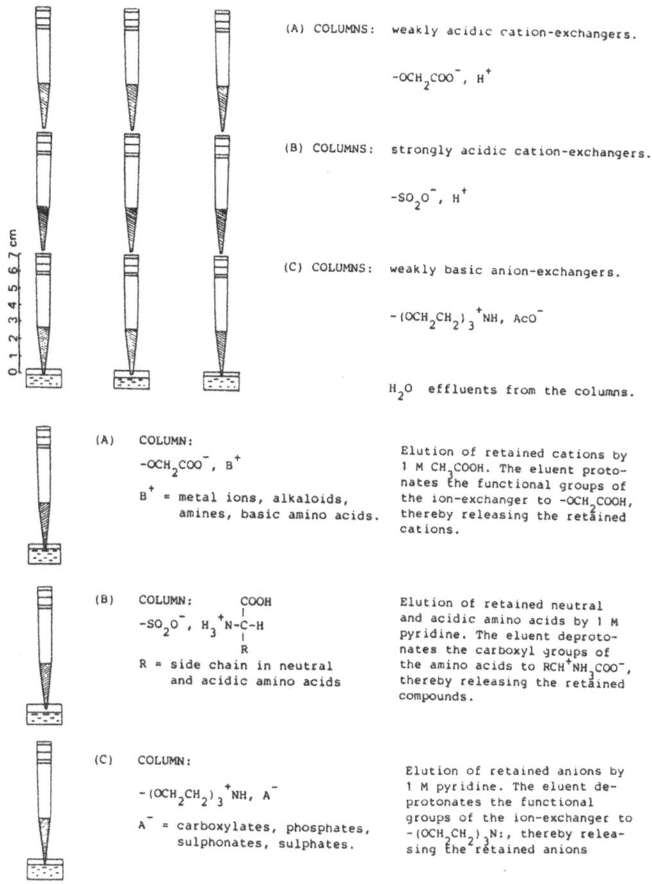

Fig.2. Columns fit up for group separation, distribution of different types of natural products on the columns after flushing with water and elution principles used for the three different types of ion-exchange columns.

Transformation of isolated glucosinolates into potassium salts

The glucosinolates isolated as pyridinium salts from the Ecteola-Cellulose ((**C**)-column; fig. 2) need to be transformed into potassium salts before further purification and separation of individual glucosinolates. For some few glucosinolates, isolation as pyridinium salts is not recommendable owing to instability of the compounds during performance of this isolation technique (vide infra).

Glucosinolates dissolved in water are transformed into potassium salts by flushing the solution through cation-exchangers as column (B)(Fig. 2), but with an appreciable size and in K^+-form (instead of H^+-form). The column is then eluted with water, and the glucosinolates in the effluent are detected by use of Uvicord (at 235 nm), spot tests with silver nitrate (Olsen and Sørensen, 1980) or HPLC analysis.

Separation and purification of individual glucosinolates

Glucosinolates isolated as potassium salts (vide supra) need generally additional purification and separations. For minor amounts, preparative HVE, PC (Olsen and Sørensen, 1980 and 1981) or preparative HPLC (vide infra) can be used. Column chromatographic techniques by use of Sephadex G10 or Sephadex G25 (fine), Polyclar AT or SPE-techniques (vide infra) are recommended for purification and separation of appreciable amounts.

Crystallisation and recrystallisation are recommendable as the last purification steps. It is not possible to give a general recommendation for crystallisation conditions but concentrated/saturated aqueous or aqueous-ethanol solutions are quite often suitable.

The final products are as other glycosides and salts often hygroscopic and need to be dried and stored in vacuo. Some glucosinolates which are especially instable and sensitive to oxidation, e.g., 4-hydroxyglucobrassicin and cinnamoylderivatives (Table 1) need in addition to be stored refrigerated in the dark.

Purity and identity of glucosinolates

It is difficult to obtain 100% pure glucosinolates. Identity of the glucosinolates can be established by a number of methods including PC, HVE, GLC, HPLC, ^1H- and ^{13}C-NMR (Olsen and Sørensen, 1981) and GLC-MS (Christensen et al., 1982). However, these methods do not give any information on the content of water, inorganic salts and some other impurities. Therefore, several other of each other independent quantitative methods of analysis have to be used for determination of the purity, e.g. UV, thymol/H_2SO_4 and myrosinase/glucose methods (vide infra).

RESULTS AND DISCUSSION

The isolation of glucosinolates has been carried out by use of the methods described in materials and methods. Thereby, it has been possible to isolate most of the glucosinolates mentioned in

Table 1 and several other glucosinolates described in previous papers (vide supra). In total we have isolated about half of the ca. 100 known glucosinolates. The results obtained from these experiments have also shown, that some glucosinolates, owing to their instability, are difficult or impossible to isolate by these methods even as strongly acidic and basic conditions are avoided. The problems have been encountered with 4-hydroxyglucobrassicin, described some years ago as an unknown hydroxyindolylglucosinolate occurring in oilseed rape (Olsen and Sørensen, 1980), and with some o-dihydroxyphenyl substituted glucosinolates as well as cinnamoylderivatives (Fig. 1 and Table 1). Change from elution of the Ecteola-Cellulose with 1 M pyridine to the eluent now applied (Bjerg and Sørensen, 1986), and use of SPE-techniques or Polyclar AT have resulted in a solution of these problems. For all of the column chromatographic methods, the glucosinolate-containing fractions are found by the methods mentioned above in the section "Transformation of isolated glucosinolates into potassium salts".

Sephadex G10 and Sephadex G25 (fine) prepared and used in columns (2.5 x 100 cm or 0.9 x 60 cm) according to the manufactors recommendations are simple and gentle methods. The columns are eluted with water and fractions of e.g. 10 ml are collected. Thereby it is possible to obtain purification of glucosinolates from some impurities, but only some few glucosinolates can be separated efficiently on these columns. Aromatic glucosinolates with strong adsorption properties are partially separated from aliphatic glucosinolates in the same way as described previously for their separation on Ecteola-Cellulose columns (Bille et al., 1983b).

Polyclar AT swollen in water, packed in columns (2.5 x 100 cm or 0.9 x 60 cm), deactivated with methanol-water (7:3) and finally washed with water gives a simple and efficient column for purification of glucosinolates from several impurities. The columns are eluted with water and water with increasing concentrations of methanol to the deactivation concentration. Fractions of 10 ml are collected. In addition, these types of columns allow separation of several glucosinolates, especially phenolic compounds.

Solid Phase Extraction (SPE-techniques) is a very efficient chromatographic separation method, which appears to be of great value in purification and separation of glucosinolates. The column material used in SPE-techniques are identical with the column material used for reversed phase HPLC, and ion-pairing, normal phase, ion-exchange as well as adsorption HPLC-materials may also be used. The reversed phase C-18 material has been used deactivated with methanol in columns (1.5 x 30 cm), resulting in starting conditions as shown in Fig. 3.

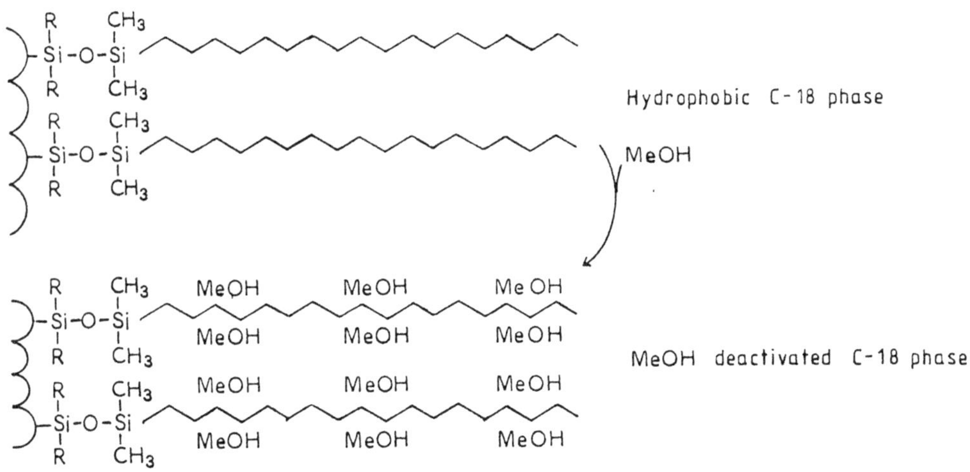

Fig.3. Illustration of the difference between normally hydrophobic C-18 phase and the C-18 phase deactivated with methanol (MeOH).

The hydrophobic C-18 phase has a hydrophobic barrier with low affinity for water and hydrophilic compounds as glucosinolates. Treatment of the column material with methanol or methanol-water (7:3) and removal of excess methanol by washing with water leaves a MeOH-deactivated column (Fig. 3). Thereby, the hydrophobic barrier is changed and it is now possible for hydrophilic compounds as glucosinolates to be bound to the stationary phase. Elution of the column is performed with water followed by mixtures of methanol-water with increasing concentration of methanol. Compounds or impurities strongly bound to the column material can be removed by use of solvents with increasing nonpolarity, e.g. water-methanol

-isopropanol-acetonitrile-acetone.

Glucosinolates are separated in this SPE-technique mainly according to their separation in analytical HPLC of intact glucosinolates (Bjerg and Sørensen, 1986), even as counter ions and buffer solutions are avoided. However, the sequence of elution of some glucosinolates are changed if we compare with reversed phase ion-pair HPLC.

Isolation and separation of minor amounts of intact glucosinolates have been performed by preparative HPLC on analytical columns. The buffer ions are not required, whereas counterions are included in the mobile phases, which otherwise are identical with those used in the analytical techniques (Bjerg and Sørensen, 1986). The counterions in the glucosinolate-containing eluate are removed by column chromatography as described in the section "Transformation of isolated glucosinolates into potassium salts" (vide supra).

Fig.4 shows HPLC separations of the isoferuloyl-derivatives of glucobarbarin (Table 1). Using the desulfoglucosinolate technique, it was possible to isolate some of the desulfoglucosinolates, but only a part of the glucosinolate amount transferred to the column was reisolated. Fig. 5 shows HPLC separations of sinapoyl-derivatives of glucoraphenin using isocratic conditions at different temperatures and detection wavelengths. As for the isoferuloyl derivatives it was not possible to have quantitative determinations based on the desulfoglucosinolate technique. These glucosinolates were more or less degraded as revealed from fig.5. It was also found in the experiments with isolation of appreciable amounts using the described column chromatographic techniques (vide supra) that it is quite unstable compounds. An efficient isolation of such compounds as well as of 4-hydroxyglucobrassicin is performed by using a combination of the Ecteola-Cellulose and the SPE technique. Cinnamoyl-derivatives of glucosinolates are especially strongly bound to the column materials, including the SPE columns (vide supra). In accordance with the theory for this technique (Bjerg and Sørensen, 1986) there are strongly effects of increased temperatures and changes in the mobile phases as well as in the hydrophobic barrier (Fig.3).

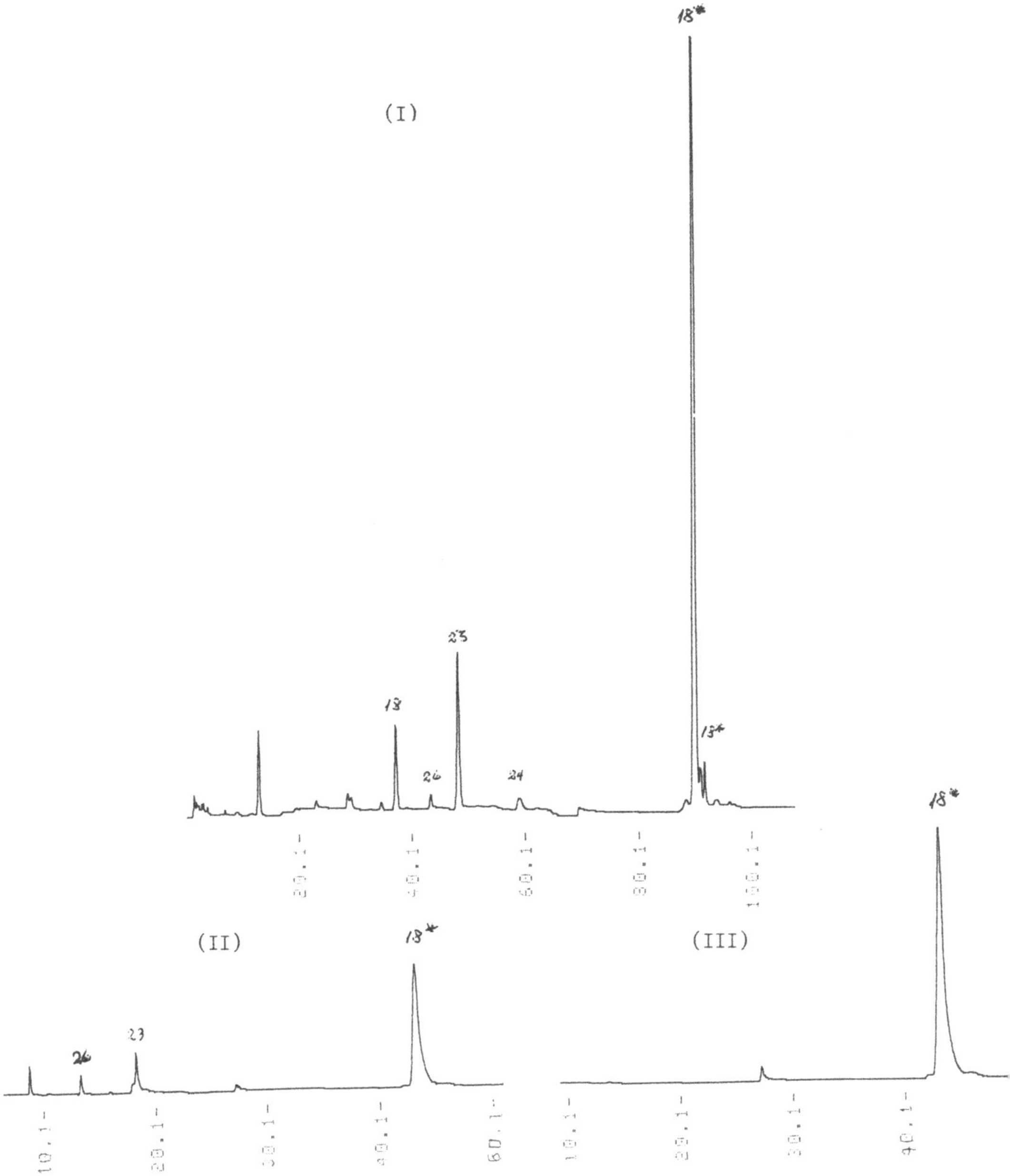

Fig.4. Separation of intact isoferuloyl-glucosinolates using gradient HPLC on column A, 50°C and detection at 235 nm (I). The numbers refer to glucosinolates shown in Table 1. In (II) and (III) are shown the separation of corresponding desulfoglucosinolates using column C, 30°C, and detection at 280 nm and 330 nm respectively. For the HPLC columns see Bjerg and Sørensen, 1986.

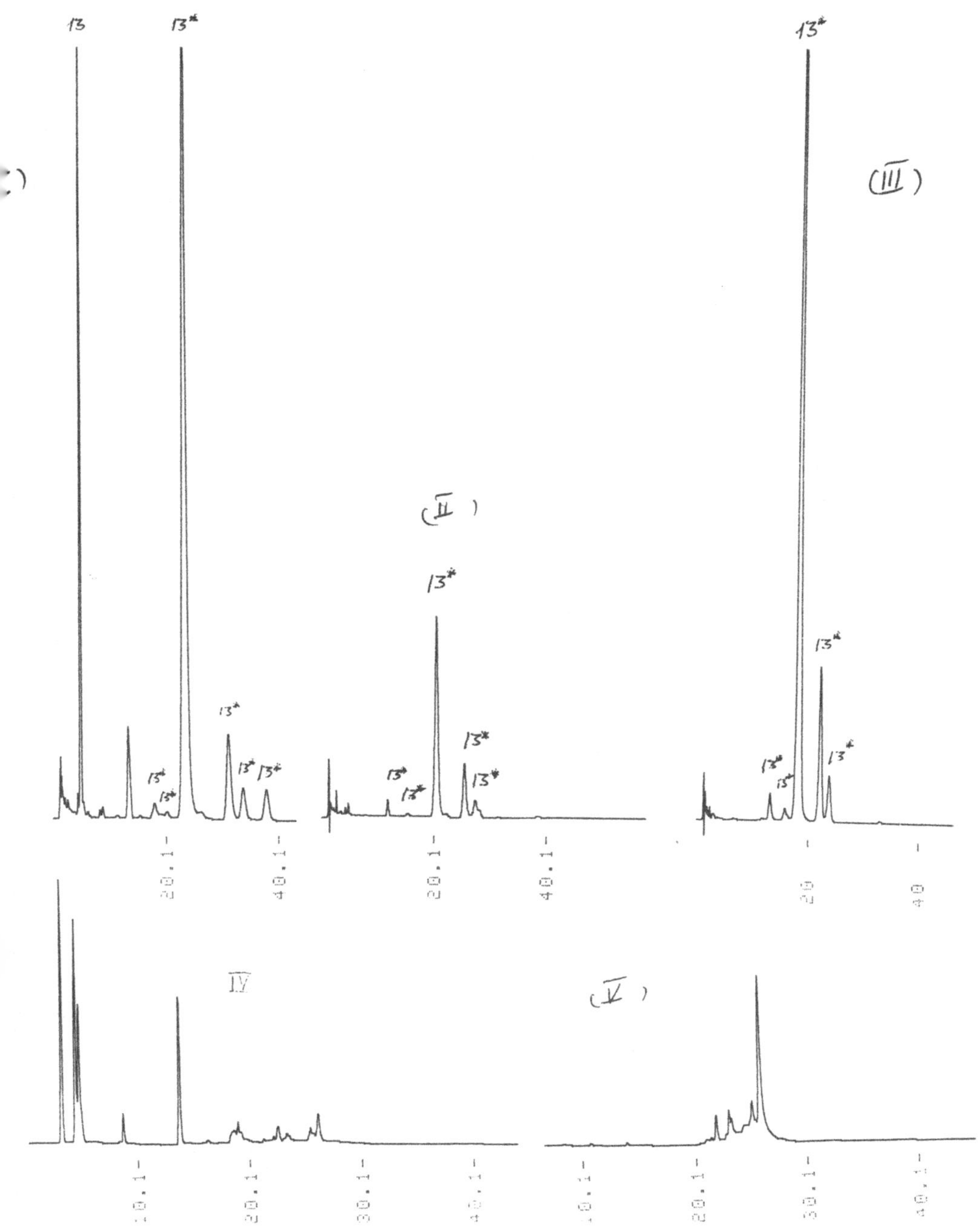

Fig. 5. Separation of intact sinapoylglucosinolates using isocratic HPLC on column A, CH_3CN (30%); 30°C and detection at 235 nm (I); 70°C and detection at 280 nm (II); 70°C and detection at 330 nm (III). The numbers refer to glucosinolates shown in Table 1. In (IV) and (V) are shown the corresponding desulfoglucosinolates separated in column C at 30°C and with 280 nm and 330 nm detection, respectively.

Isoferuloylglucosinolates

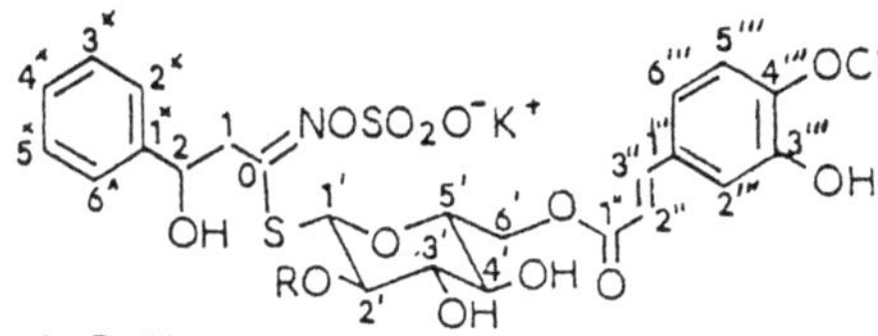

1; R=H

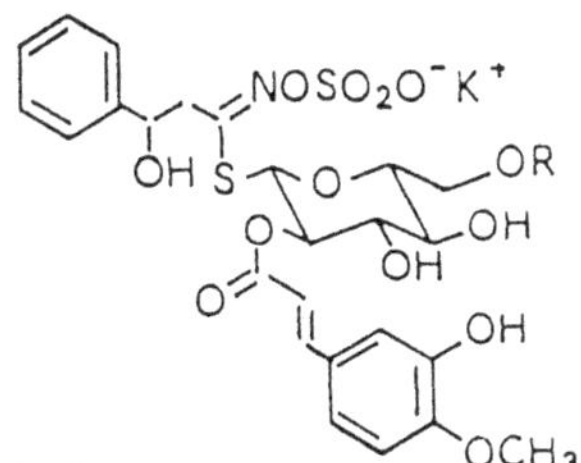

2; R=H

3; R=Isoferuloyl

Sinapoylglucosinolates

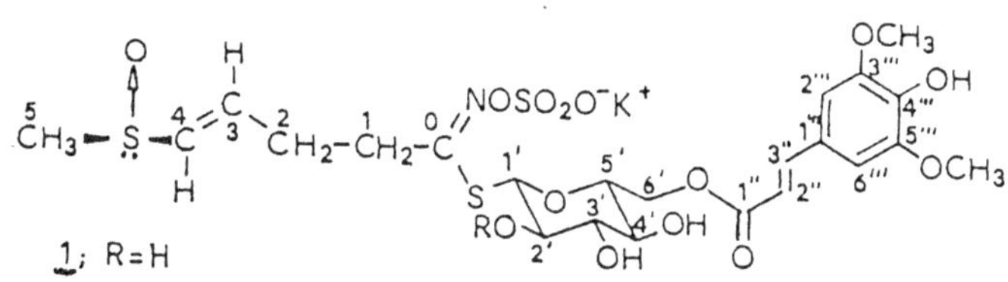

1; R=H

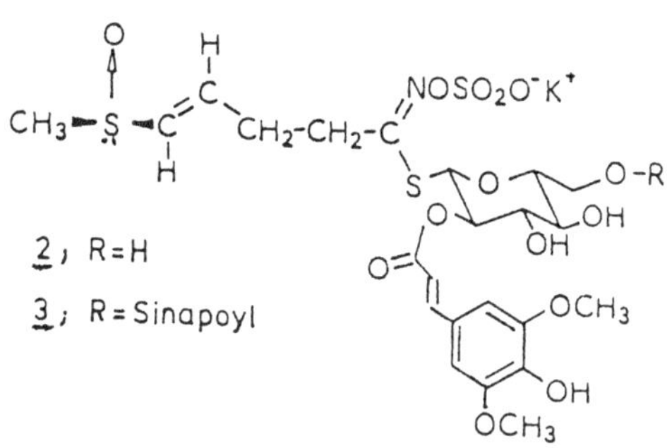

2; R=H

3; R=Sinapoyl

4 ; R_2=H; R_6=Sinapoyl

5 ; R_2= Sinapoyl; R_6=H

6 ; R_2=R_6= Sinapoyl

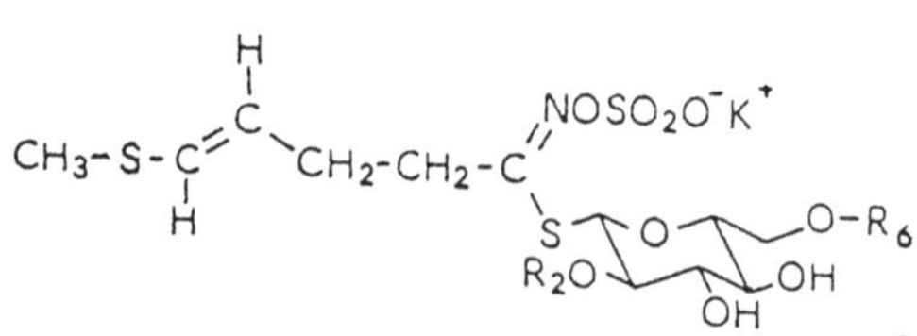

Fig. 6. Structures of some of the identified cinnamoyl-glucosinolates

Determination of the purity of isolated intact glucosinolates includes methods mentioned in the materials and methods. In addition to these methods, quantitative UV determinations, quantitative determinations of glucose released by myrosinase treatment (Heaney and Fenwick, 1981), quantitative determination of the colour produced in the thymol-sulphuric acid method (Brzezinski and Mendelewski, 1984) and spectrophotometric determination of Pd-glucosinolate complexes (Møller et al., 1985b) have been used. The results have revealed, that it is necessary to use all of these methods, as no one of them gives reliable results for all of the glucosinolates. Correct determinations of the purity of glucosinolates used as internal or external references for quantitative methods of glucosinolate analysis are required. The same holds for glucosinolates used for determination of response factors and for development of correct and reliable analytical methods. Details of the methods and results from these experiments will be presented elsewhere.

CONCLUSION

The methods now available for isolation of intact glucosinolates allow isolation of appreciable amounts of the individual compounds. Simple methods are available for isolation of glucosinolates for reference purposes. The new techniques (SPE) give also an important fundament for those working with methods of glucosinolate analysis and it allows separation of both unstable glucosinolates and complex mixtures of glucosinolates.

ACKNOWLEDGEMENTS

Support from the Danish Agricultural Research Council and from CEC is gratefully acknowledged.

REFERENCES

Andersen, H.R. and Sørensen, H. (1985) Double low rapeseed meal in diets to young bulls. In: Advances in the Production and Utilization of Cruciferous Crops (Ed. H. Sørensen) Martinus Nijhoff Publ. Dordrecht, pp. 208-217.

Bille, N., Eggum, B.O., Jacobsen, I., Olsen, O. and Sørensen, H. (1983a) The effects of processing on antinutritional rape constituents and the nutritive value of double low rapeseed meal. Zeitschr. Tierphysiol., Tierernährung u. Futtermittelkd. 40, 148-163.
Bille, N., Eggum, B.O., Jacobsen, I., Olsen, O. and Sørensen, H. (1983b) Antinutritional and toxic effects in rats of individual glucosinolates (±myrosinases) added to a standard diet. 1.The effects on protein utilization and organ weights. ibid. 49, 195-210.
Bjerg, B., Olsen, O., Rasmussen, K.W. and Sørensen, H. (1984) New principles of ion-exchange techniques suitable to sample preparation and group separation of natural products prior to liquid chromatography. J. Liquid Chromatogr. 7, 691-707.
Bjerg, B., Fenwick, G.R., Spinks, A. and Sørensen, H. ((1986a) Failure to detect glucosinolates in cocoa. Phytochemistry (submitted).
Bjerg, B., Eggum, B.O., Jacobsen, I., Jensen, J., Larsen, L.M. and Sørensen, H. (1986b) Ernæringsmæssige og toksiske effekter af seks individuelle glucosinolater iblandet en glucosinolatfri diæt og fodret til rotter. Statens Husdyrbrugsforsøg, Meddelelse Nr. 619, 1-4.
Bjerg, B. and Sørensen, H. (1986) Quantitative analysis of glucosinolates in oilseed rape based on HPLC of desulfoglucosinolates and HPLC of intact glucosinolates (This volume).
Brzezinski, W. and Mendelewski, P. (1984) Determination of total glucosinolate content in rapeseed meal with thymol reagent. Z. Pflanzenzüchtg. 93, 177-183.
Buchwaldt, L., Nielsen, J.K. and Sørensen, H. (1985) Preliminary investigations on the effect of sinigrin on in vitro growth of three fungal pathogens of oilseed rape. In: Advances in the Production and Utilization of Cruciferous Crops (Ed. Sørensen, H.) Martinus Nijhoff Publ., Dordrecht pp. 73-84.
Buchwaldt, L., Larsen, L.M., Plöger, A. and Sørensen, H. (1986) FPLC isolation and characterization of plant myrosinase, β-thioglucoside glucohydrolase, isoenzymes. J. Chromatogr. 363 (1), 71-80.
Christensen, B.W., Kjær, A., Madsen, J.-Ø., Olsen, C.E., Olsen, O. and Sørensen, H. (1982) Mass-spectrometric characteristics of some pertrimethylsilylated desulfoglucosinolates. Tetrahedon 38, 353-357.
Dahlgreen, R. (1975) Bot.Not. 128, 119.
Eggum, B.O., Larsen, L.M., Poulsen, M.H. and Sørensen, H.(1985a) Conclusions and recommendations. In: Advances in the Production and Utilization of Cruciferous Crops (Ed. Sørensen, H.) Martinus Nijhoff Publ., Dordrecht pp. 73-84.
Eggum, B.O., Olsen, O. and Sørensen, H. (1985b) Effects of glucosinolates on the nutritive value of rapeseed. ibid. pp.50-60.
Eggum, B.O., Just, A. and Sørensen, H. (1985c) Double low rapeseed meal in diets to growing - finishing pigs. ibid. pp. 167-176.
Heaney, R.K. and Fenwick, G.R. (1981) A micro-column method for the rapid determination of total glucosinolate content of cruciferous material. Z. Pflanzenzüchtg. 87, 89-95.
Larsen, L.M., Olsen, O. and Sørensen, H. (1983) Failure to detect glucosinolates in Plantago species. Phytochemistry 22,2314-2315

Larsen, L.M., Olsen, O., Pedersen, L.H. and Sørensen, H. (1984) N^5-(4-hydroxybenzyl)glutamine, 4-hydroxybenzylamine and 4-hydroxybenzylglucosinolate in Sinapis species. Phytochemistry 23, 895-896.
Larsen, L.M., Nielsen, J.K., Plöger, A. and Sørensen, H. (1985) Responses of some beetle species to varieties of oilseed rape and to pure glucosinolates. In: Advances in the Production and Utilization of Cruciferous Crops (Ed. Sørensen, H.) Martinus Nijhoff Publ., Dordrecht pp. 230-244.
Møller, P., Olsen, O., Plöger, A., Rasmussen, K.W. and Sørensen,H. (1985a)Quantitative analysis of individual glucosinolates in double low rapeseed by HPLC of intact glucosinolates. ibid. pp. 111-126.
Møller, P., Plöger, A. and Sørensen, H. (1985b) Quantitative analysis of total glucosinolate content in concentrated extracts from double low rapeseed by the Pd-glucosinolate complex method ibid. pp. 97-110.
Olsen, O. and Sørensen, H. (1979) Isolation of glucosinolates and the identification of o-(α-L-rhamnopyranosyloxy) benzylglucosinolate from Reseda odorata. Phytochemistry 18, 1547-1552.
Olsen, O. and Sørensen, H. (1980) Sinalbin and other glucosinolates in seeds of double low rape species and Brassica napus cv. Bronowski. J. Agric. Food Chem. 28, 43-48.
Olsen, O. and Sørensen, H. (1981) Recent advances in the analysis of glucosinolates. J. Am. Oil Chem. Soc. 58, 857-865.
Rodman, J.R. (1981) In: Phytochemistry and Angiosperm Phylogeny (Eds. Young, D.A. and Siegler, D.S.). Praeger, New York, p. 43.
Sørensen, H. (1985a) Advances in the Production and Utilization of Cruciferous Crops (Ed. Sørensen, H.) Martinus Nijhoff Publ., Dordrecht.
Sørensen, H. (1985) Kvalitet af raps og muligheder for optimal udnyttelse til minkfoder via udnyttelse af viden fra forsøg med rapsfodring af forskellige dyr. In: Proceedings fra NJF--seminarium nr. 85 i Ålborg, September 1985. Husdyrsektionen sektion V, Subsektionen for pelsdyr.

COMPARISON OF PROCEDURES FOR OPTIMUM EXTRACTION. PURIFICATION AND ANALYSIS OF DESULFO INDOLYL GLUCOSINOLATES

R. Buchner

Institute for Agronomy and Plant Breeding
University of Göttingen
von Siebold Strasse 8
D-3400 Göttingen
Federal Republic of Germany

ABSTRACT

Using HPLC instead of GLC and thus including indolyl glucosinolates in glucosinolate analysis requires the development of new methods for determination of glucosinolates in this case derived from the procedures used for the desulfo alkenyl glucosinolates. Efforts are reported, to create a common framework for glucosinolates' analysis for different purposes (official method, screening method; intact and desulfo glucosinolates' method). Procedures using extraction solutions of different pH respectively containing antioxidants are tested for their effect on stability of glucosinolates. As a result the use of an acetate buffer pH 4.0 at 39°C is recommended for the desulfation step. A number of ion exchange materials were tested for their suitability for purification of glucosinolates. Some observations of behaviour of glucosinolates during extraction, storage and HPLC analysis are reported.

INTRODUCTION

Sang and Truscott (1984) were the first who tried to stop the loss especially of 4-hydroxy-glucobrassicin (4OHGBC), which was observed in many laboratories working on desulfo glucosinolates. They introduced the use of 2-mercaptoethanol at pH 8 and found a twofold increase of 4OHGBC. However the use of this antioxidans is not possible for a greater amount of samples because of its smell and toxicity.

There is only little knowledge about maximum yield and stability of glucosinolates especially also of 4OHGBC. Certainly the most sensitive step during sample preparation is the desulfation step.

Therefore the attempt was made, to find optimal conditions for the desulfation step (introduced by Thies, 1979) and afterwards to transfer the experiences to the other parts of sample treatment.

During all the work the creation of a common framework for intact (Møller et al., 1985) and desulfo glucosinolates' analysis, for most precise and screening analyses was an idea in the background.

MATERIALS AND METHODS

Apparatus

(a) Pasteur pipettes, glass wool placed in the tips. as ion exchange columns.

(b) Water bath or heating plate.

(c) Small laboratory centrifuge.

(d) Millipore dip-in-filter type Immersible CX-10.

(e) Low temperature oven.

(f) UV spectrophotometer.

(g) HPLC chromatograph: dual pump gradient instrument with a UV detector (two Waters M510 pumps, Waters WISP M710b. Data Module M730, System Controller M721 and UV detector M441 with the atomic emission light of a cadmium lamp (229 nm) as source).

(h) LC column - A C18 reversed phase column Waters Nova-pak 150 x 3.9 mm, 4 um particle size.

Reagents

(a) 15 mmol/l glucotropaeolin, 15 mmol/l sinigrin (both own preparation, nearly free from foreign glucosinolates and of glucose-test (Heaney and Fenwick, 1981) defined purity) and 3 mmol/l sulfanilic acid as internal standards.

(b) 20 mM/l Tris buffer pH 7-8, 20 mM/l sodium carbonate buffer pH 6.5-7, 20mM/l sodium acetate buffer pH 4.

(c) 20 mM/l potassium phosphate buffer pH 4.

(d) Anion exchangers: DEAE-Sephadex A-25 (Pharmacia Fine Chemicals, Uppsala), Servacel Ecteola 23, Serdolit AW-2, AW-4, AW-14, Duolite A 30 B (all Serva, Heidelberg). Swollen in 2 M acetic acid, filled in the pasteur pipettes (Ecteola: plastic pipette tips), rinsed with the 10 fold volume 6 M imidazole formiate, rinsed again with the 20 fold volume of water.

(e) Sulfatase type H-1 (Sigma, St. Louis).

(f) DEAE-Sepharose C1-6B formiate form (Pharmacia Fine Chemicals, Uppsala).

(g) 0.2 mol/1 sodium acetate.

(h) ultra pure water and acetonitrile for HPLC.

Purification of the sulfatase

(1) 200 mg sulfatase solved in 20 ml water.

(2) passed through 20 ml Sepharose column (diameter 26 mm).

(3) rinse with 40 ml water.

(4) elution with 60 ml 0.2 mol/1 sodium acetate.

(5) concentration of eluate to ca. 4 x 100 ul using four dip-in-filter.

(6) dilution to 4 x 2 ml with water.

(7) concentration to 4 x 100 ul once more.

(8) Rinse the filters, combine and dilute the filtrates to 20 ml sulfatase stem solution.

(9) Store freezed.

(10) Regeneration of column: rinse with 10 ml 6 M/1 imidazole formiate and 30 ml water subsequently (one column filling can be used up to 10 times).

(11) 50 ul of sulfatase stem solution + 2 ml 0.15 mmol/1 sinigrin (stem solution 1:10 diluted) should result at least in an absorption decrease at 227 nm of 0.0338 absorption units/minute.

Sample preparation for glucosinolate analysis as it is used in our laboratory now

(1) Weigh in 0.2 g of grinded seed.

(2) Add 1.5 ml hot 70% methanol/water mixture and immediately afterwards 200 ul of internal standard glucotropaeolin (1 part standard to 2 parts final volume diluted for 00-rapeseed). Shake.

(3) Extract 10 minutes at 80°C.

(4) Centrifuge 3 minutes at 3000 rpm.

(5) Decant supernatant and repeat step (2) to (5) for the precipitate adding more internal standard two times.

(6) Combine supernatants and fill up to 5 ml.

(7) Put 500 to 1000 ul of the extract to 20 mg Sephadex columns.

(8) Rinse two times with 1 ml 20 mmol/l sodium acetate buffer pH 4.0.

(9) Add 75 ul sulfatase solution (1 part sulfatase stem solution + 1/2 part water)

(10) Cap the columns.

(11) Desulfation 6 hours at 39°C.

(12) Elution with 1.5 ml water. Shake the samples.

(13) If necessary filter the samples.

(14) Injection of 30-80 ul into HPLC. Used program: 0-16.5 min 0-16.5% acetonitrile, 16.5-18 min 16.5-0% acetonitrile (rest: water). Flow speed: 2.0 ml/min.

(15) Detection at 229 nm.

(16) Calculation of amounts glucosinolates using the area of the internal standard and the relative response factors reported elsewhere (Buchner and Thies, this volume).

The following glucosinolates were examined (in order as they occur in the HPLC chromatogram at room temperature): glucoiberin, progoitrin, sinigrin, gluconapoleiferin, an unknown benzyl-type glucosinolate, gluconapin, 4-hydroxy-glucobrassicin, glucobrassicanapin, glucotropaeolin, glucobrassicin, gluconasturtiin and neoglucobrassicin.

For the presented work different steps were varied. If done so, it is indicated in the text.

RESULTS

Test of ion exchangers

On principle there exist three types of frameworks for ion exchangers. Frameworks derived from aromatic compounds e.g. styrol (1), those derived form amides e.g. acrylamide (2) and those derived from saccharides e.g. cellulose (3). The first group was excluded from the start because of feared adsorption of aromatic glucosinolate side chains to the exchanger.

After having tested different ion exchangers of the second group (Serdolit AW-2, AW-14 and Duolite A 30 B) this group was also excluded because of more or less retention of 4OHGBC. An interaction between the hydroxy-group of the indolyl moiety and the amide-group of the exchanger is guessed. Glucobrassicin and neoglucobrassicin are eluted quantitatively.

The ion exchanger of the third group remain. Ecteola, which is a condensation product from epichlorhydrin, cellulose and triethanolamin (!) is not proved yet to retain 4OHGBC. This is, because its low capacity makes it impossible, to use Ecteola for the desulfo method.

Comparing DEAE Sepharose andDEAE Sephadex the already earlier used Sephadex showed the best properties (especially higher capacity). The use of 50 mg columns instead of 20 mg columns didn't result in a change of the composition of glucosinolates, which gives some evidence, that there isn't any unspecific adsorption of certain moieties of glucosinolates neither to the framework nor to the active groups (tertiary nitrogen) of DEAE Sephadex A-25 (see also fig. 2).

The idea of a common ion exchanger for the desulfo and the intact glucosinolates' method couldn't be realised yet, as there wasn't any exchanger found with a low pK value for the elution of glucosinolates in the intact method, a polysaccharide framework and a high capacity.

DEAE Sephadex A-25 tolerated 70% methanol/water solutions. It looses about 1/3 of its capacity, which is however still high enough for a sample containing 200 umol glucosinolates per g defatted meal.

Therefore we recommend the use of 20 mg DEAE Sephadex A-25 columns for the preparation of desulfo glucosinolates.

Figure 1 and 2 show a characterization of the Sephadex ion exchange material. Although a retention of 4OHGBC couldn't be found, this compound evidently shows the slowest elution from the exchanger.

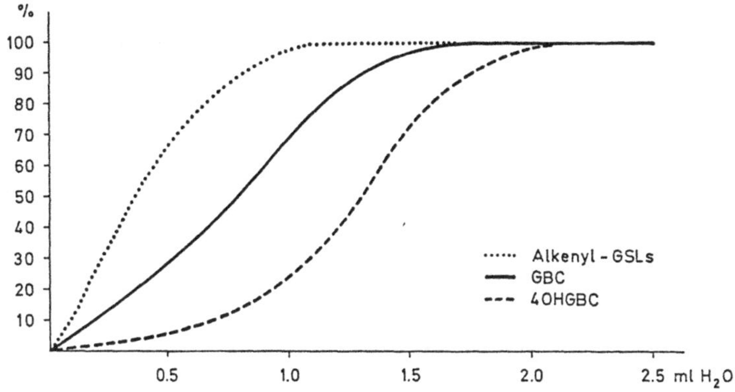

Fig. 1 Elution profiles of the different glucosinolates from a 80 mg DEAE Sephadex A-25 column.

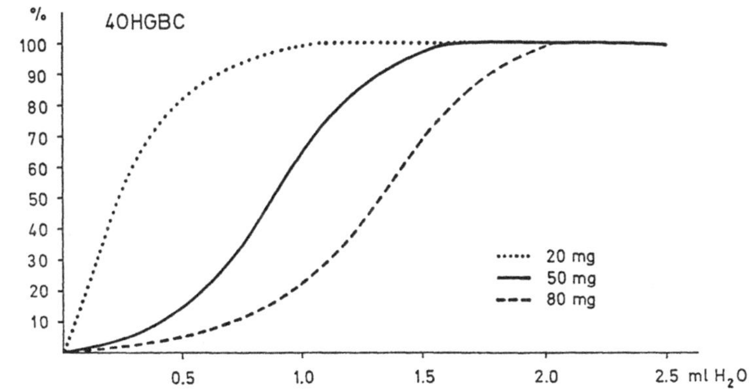

Fig. 2 Elution profiles of 4OHGBC from different amounts of DEAE Sephadex A-25.

Optimisation of the on-column desulfation step

The use of 2-mercaptoethanol only in the unbuffered washing solution used after having put the samples onto the columns caused an increase in the yield of 4OHGBC of about 20 - 30%. As the smell and the toxicity of this antioxidans isn't acceptable for a standard method, a search for another antioxidans was performed. L-cystein, ascorbic acid, dithioerythritol, dithiothreitol and glutathione were tested in different concentrations without success. Either the protection of 4OHGBC was incomplete or the enzyme activity was diminished or both.

Therefore it was preferred, to attempt the use of buffers of different pH and strongness and different temperatures during the desulfation. To get a more complete knowledge about the desulfation process and the maximum yield of glucosinolates the time dependency of the desulfation of the different glucosinolates was studied.

Influence of different pH buffers

Buffers were only used for washing the ion exchanger columns after the adsorption of the sample. A concentration of 20 mmol/l was found to be suitable. Below 5 mmol/l the protective effect of the buffer was gone. At concentrations higher than 30 mmol/l the velocity of the desulfation may be reduced.

The choice of the buffer may influence the action of the enzyme. For example the use of Tris buffer at pH 7 gives the twofold yield of 4OHGBC compared to a sodium carbonate buffer of identical strongness.

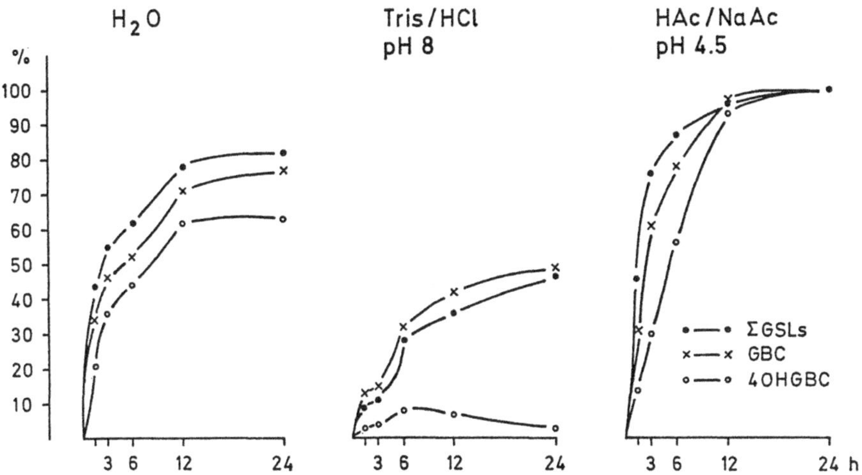

Fig. 3 Desulfation kinetics and yield of different typical glucosinolates and the total amount glucosinolates at different pH. The pH of the water was 4.5-5. The yield after 24 hours desulfation at pH 4.5 is set to 100%. 75 ul of 1:5 diluted H-1 were added.

Figure 3 shows six important results:

(1) Reaching a stable glucosinolate level doesn't mean automatically the complete desulfation of all the glucosinolates of a sample.

(2) A low pH raises the yield of all glucosinolates, but especially of 4OHGBC.

(3) It depents on the pH, which glucosinolate is preferred by the sulfatase. For example glucobrassicin is converted faster than the average of the glucosinolates at pH 8, what it is not when the sample was washed with water. This may become a source for problems of different glucosinolates reach their maximum yield at different pH (this isn't found yet, but is possible).

(4) The use of water and buffer of identical pH causes different answers in the desulfation process.

(5) The approach proposed by Sang and Truscott (1984) - use of pH 8 - is not useable for the preparation of desulfo gluco-

sinolates.
(6) A pH of 8 without the presence of an antioxidans causes degradation of 4OHGBC.

Influence of different temperature at different pH

The maximum yield of all glucosinolates is much more influenced by pH than by temperature.

The velocity of desulfation of glucosinolates is much more influenced by temperature than by pH.

Higher temperature create more acidic conditions if the same buffer is used at low and high temperatures. It might be 0.5 pH units from 21 to 45°C.

The yield of 4OHGBC is stronger influenced by the temperature than the yield of the other examined glucosinolates. As shown in figure 4 (full bars) the optimum yield of 4OHGBC is reached at 39°C. The position of the optimum is the same at pH 4.0 and 6.5, but the maximum amount of this glucosinolate is less at pH 6.5 than pH 4.0.

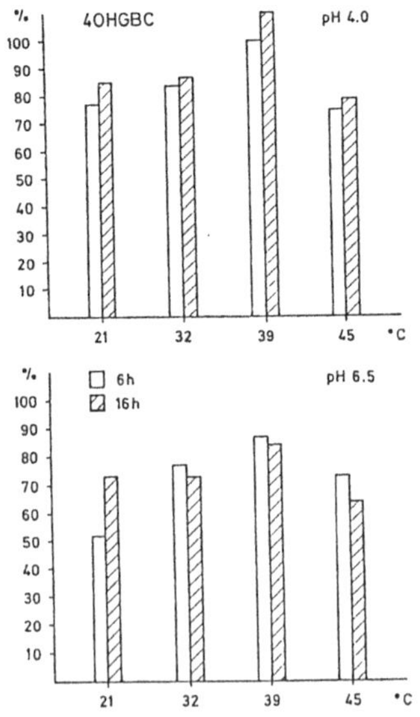

Fig. 4 Amount of 4OHGBC after 6 and 16 hours desulfation at 4 different temperatures and 2 different pH. Yield after 6 hours desulfation at 39°C and pH 4.0 is set to 100%. 75 ul of 1:1.5 diluted H-1 stem solution added.

Stablilty of glucosinolates during on-column desulfation

If a buffer of pH 4.0 or even pH 6.5 is used, the stability of 4OHGBC doesn't seem to be a problem any longer (full and single slope bars of figure 4 for desulfation times of 6 and 16 hours). The increase in 4OHGBC at 39°C and pH4 from 6 to 16 hours desulfation time was not repeatable in other experiments.

But at higher temperatures problems arise with the stability of glucotropaeolin and other benzyl- or phenyl-glucosinolates (e.g. also gluconasturtiin). As shown in figure 5 there is a loss of glucotropaeolin during desulfation, if columns were washed with water or buffers of pH higher than 4.5.

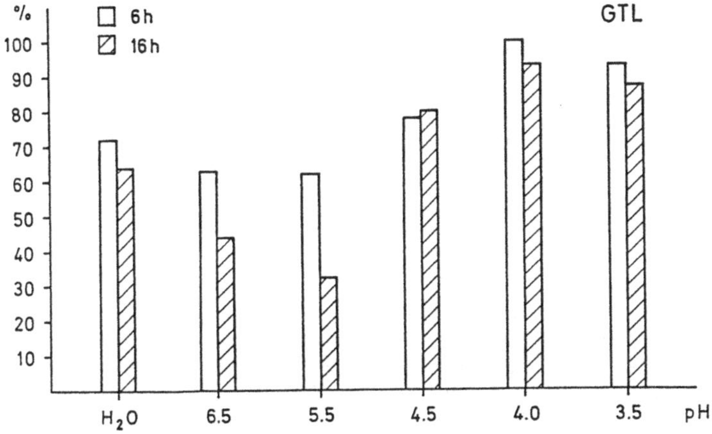

Fig. 5 Yield and stability of glucotropaeolin at 39°C at different pH. Yield after 6 hours desulfation at pH 4 is set to 100%. 75 ul of 1:1.5 diluted H-1 added.

As finally indicated in figure 6 maximum yield (more than received if 2-mercaptoethanol was added) and good stability of all glucosinolates examined were found at 39°C and pH 4.0 sodium acetate buffer, 20 mmol/l.

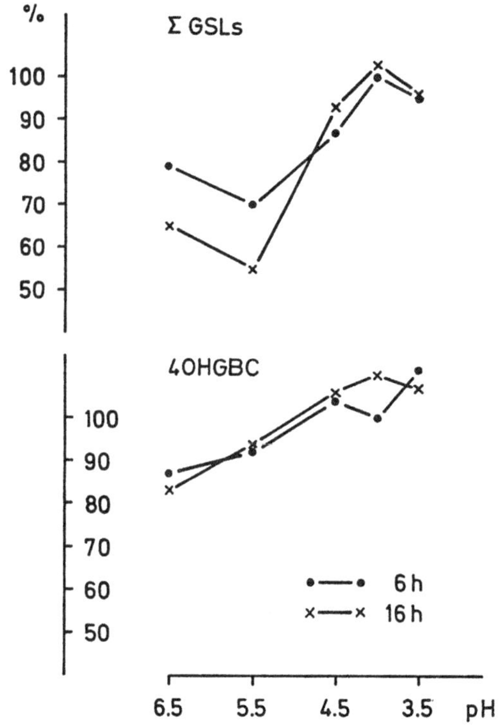

Fig. 6 Yield and stability of the total of glucosinolates and 4OHGBC at different pH at 39°C. Yield after 6 hours desulfation at pH 4.0 is set to 100%. 75 ul of 1:1.5 diluted H-1 added.

Optimisation of extraction

The creation of a common framework for screening and most precise methods is of importance for our laboratory. This is for example the reason of the fact, that we mill the seed before adding the hot extraction solutions. Therefore we had to ensure, that there isn't any myrosinase activity in the grinded rape seed.

One week old grinded seed, freshly grinded seed and in the hot extraction solution crushed seed (use of Ultra Turrax) were compared and no difference found between the different samples.

Suppositions that there is a myrosinase activity in the grinded seed (below 10% humidity!) probably are derived from experiments, where the attempt was made to extract glucosinolates using 100% methanol at room temperature. It was however found, that all examined glucosinolates but especially 4OHGBC are less soluble in methanol than they are in water.

Thus there is no myrosinase activity in the grinded seed remarkable.

The positive effect of the methanol is the break down of cell structures. In one experiment a short preincubation with a small amount of 60% methanol and afterwards adding a greater volume hot water showed a slightly better extraction than the use of 70% methanol.

Adding acidic buffers instead of water for extraction - to protect sensitive glucosinolates - retards the extraction process.

4OHGBC is not stable in unbuffered extraction solutions when stored in the refrigerator.

Storage of ready desulfoglucosinolate samples and optimisation of HPLC

4OHGBC is not stable in the solutions received by elution of the Sephadex columns with water. No loss of another glucosinolate was observed.

Figure 7 shows an effect noticed at experiments testing different HPLC column temperatures for better separation of single peaks.

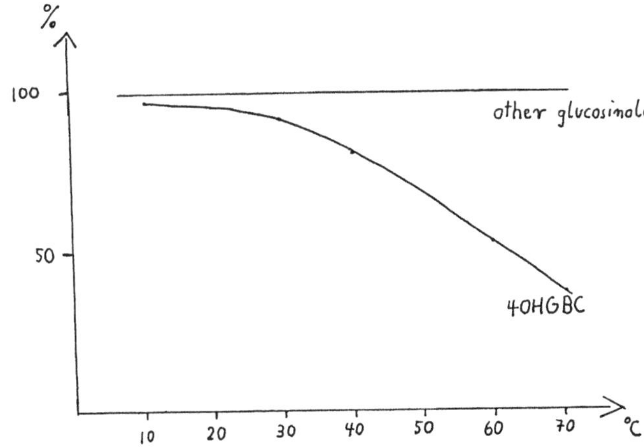

Fig. 7 One desulfo glucosinolate sample run at different temperatures. The starting point of the 4OHGBC curve at the left hand side is set at random.

The area of 4OHGBC decrease dramatically with increasing temperature. The areas of the other glucosinolates also glucobrassicin and neoglucobrassicin remain stable. A shift in the absorption spectrum of 4OHGBC is possible but certainly not the main reason.

Using a 20 mmol/l potassium phosphate buffer pH 4.0 instead of water in the HPLC resulted in an increase of the area of 4OHGBC yet also (in a lower extent) of glucobrassicin and neoglucobrassicin. The area of 4OHGBC also becomes more temperature independent. The increase of 4OHGBC through the use of the buffer depends on the age of the sample.

This indicates, that 4OHGBC isn't destroyed irreversibly at once, when a decrease in the UV absorption area is watched. Furthermore the buffer causes a stabilisation of 4OHGBC in the HPLC, but an additional shift in the absorption spectrum can't be excluded because of the higher response to the two other indolyl glucosinolates in a buffer/acetonitril system.

CONCLUSIONS

There is enough evidence to change the conditions for the desulfation step as it is proposed to pH 4.0 at 39°C for 6 hours if an equivalent to 50 ul sulfatase stem solution is added.

The principle of the use of slightly acidic conditions to protect sensitive glucosinolates will be extended to extraction, storage solutions and HPLC analysis, if appropiate. As experimental work at these parts of sample handling isn't finished yet, the steps should be kept in the established way till further results are available.

ACKNOWLEDGEMENTS

The author thanks Prof.G. Röbbelen and Prof.W. Thies for their support and helpful discussions during this work. He also thanks Miss K. Fischer and Mr.U. Ammermann for excellent technical assistance and in particular Miss Fischer for drawing the figures. Financial support of the EC Commission is greatfully acknowledged.

REFERENCES

Buchner, R., 1986. Approach to determination of HPLC response factors for glucosinolates. This volume.

Heaney, R.K., Fenwick, G.R. 1981. A micro-column method for the rapid determination of total glucosinolate content of cruciferous material. Z. Pflanzenzüchtg., 87,89-95.

Møller, P., Olsen, O., Plöger, A., Rasmussen, K.W., Sørensen, H. 1985. Quantitative analysis of individual glucosinolates in double low oilseed rape by hplc of intact glucosinolates. In "Advances in the production and utilization of cruciferous crops with special emphasis to oilseed rape" (Ed. h. Sørensen), Copenhagen, p. 111-126.

Sang, J.P., Truscott, R.J.W. 1984. Liquid chromatographic determination of glucosinolates in rapeseed as desulfoglucosinolates. J. Assoc. Off. Anal. Chem.,67. 829-833.

Spinks, E.A., Sones, K., Fenwick, G.R. 1984. The quantitative analysis of glucosinolates in cruciferous vegetables, oilseeds and forage crops using high performance liquid chromatography. Fette, Seifen, Anstrichmittel, 86.228-230.

Thies, W. 1979. Detection and utilization of a glucosinolate sulfohydrolase in the edible snail, Helix pomatia. Naturwiss., 66, 364.

OPTIMIZATION OF GLUCOSINOLATE DESULPHATION BEFORE HPL-CHROMATOGRAPHY

A. QUINSAC and D. RIBAILLIER

Centre technique interprofessionnel des oléagineux métropolitains (CETIOM), Laboratoire d'Analyses, Avenue de la Pomme de Pin, Ardon, 45160-Olivet, France.

ABSTRACT

The aryl-sulphate sulphohydrolase enzyme (EC 3.1.6.1.) extracted from H. Pomatia is now quite commonly used in the analysis of glucosinolates. It gives a satisfactory sample-purification and provides derivatives which are more easy to determine by HPLC or to silylate for a gas chromatography analysis.

Its activity depends on certain factors : temperature, pH, nature of the substrate, presence of a chemical agent, the action of which must be known and controlled.

As the glucosinolate fixation on the ion exchanger reduces the sulphatase activity appreciably, higher yields of desulphoglucosinolates were obtained from extracts of non-purified glucosinolates, in which enzymes and substrates were in solution. In particular, this technique should give us the possibility to determine all the 4-OH-glucobrassicine and to define the optimal conditions of desulphation on ion exchangers.

Added to the results obtained elsewhere for the different analytical stages, these results should allow us to work out a reliable and reproducible method.

INTRODUCTION

The analysis of glucosinolates is now moving towards HPLC. This more sensitive technique gives the possibility to determine a great number of glucosinolates without initiating enzymatic degradations or drastic derivations. Such reactions are important sources of undesirable errors in reference methods. That is why glucosinolates are analysed after being purified in their intact (1. Moller and al. 1984) or desulphated (2. Spinks and al. 1984) forms.

The aryl-sulphate sulphohydrolase enzyme (EC 3.1.6.1.) is used to produce desulphated derivatives. Its action was defined and applied by THIES (3. Thies.1979 and 4. Thies. 1978). Although its behaviour towards alkenyl-glucosinolates has been studied (5. Quinsac et al.1984) and does not raise many difficulties, the determination of indolglucosinolates, and particularly of 4-OH-glucobrassicine, implies certain precautions (6. Sang et al.1984).

Therefore, we tried to find the best operational conditions leading to maximum yields of desulphated glucosinolates and particularly of 4-OH-desulphoglucobrassicine.

2. MATERIAL AND PRODUCTS

2.1. MATERIAL

- an HPLC-device with :
  . a binary gradient system KNAUER 6400 under high pressure,
  . a Rheodyne-valve with a 20 µl-injection loop,
  . a UV-spectrophotometer KNAUER 8700
  . a column 250 mm x 4,6 mm Spherisorb ODS 2 5µ
  . a pre-column 30 mm x 4,6 mm Spherisorb ODS 2 5µ.
- an integrator recorder Shimadzu CR-3A ;
- ion-exchange micro-columns :
  PIERCE disposable columns 6ml, inner diameter 8 mm, filled with 50 mg of DEAE Sephadex A25. Regeneration with 2ml of imidazol formate 6M. Double washing with 2ml/water.
- a UV-spectrophotometer SAFAS 2000 D
  equipped with quartz cuvettes and a temperature regulator of the cuvette-carrier ;
- immersible filters Millipore PTGC 11 K 25 ;
- a centrifuge/5000 revolutions per minute MLW T5 ;
- a screw-crusher Schütt, auto-cleaning ;
- an analytical balance ;
- a water bath adjustable at 95°C ;
- an agitator Vortex ;
- centrifuge tubes 10 and 30 ml ;
- filters 0,22 µ with a small dead volume ;
- cartridges Norganic Millipore ;

2.2. PRODUCTS

- acetonitrile for HPLC ;
- water prepared for the cartridges Norganic ;

- sinigrin (Aldrich) solution 2mM ;
- barium and lead acetate 0,5 M ;
- imidazole formate 6 M ;
- pyridine acetate 0,02 M ;
- sodium acetate 0,2 M ;
- ethanol ;
- sulphatase type H1 Sigma (S. 9626) ;
- DEAE Sepharose Cℓ 6B ;
- SP Sephadex C25 ;
- Desulphation solution A
  . ethylene diamine buffer, acetic acid pH 5,8 20 mM ;
  . EDTA 0,1 mM ;
  . 2-mercaptoethanol 5 mM ;
- Desulphation solution B
  . buffer TRIS 18 mM pH 8,0 ;
  . EDTA 0,1 mM ;
  . 2-mercaptoethanol 5 mM ;
- Extraction solution
  . EDTA 1 mM ;
  . 2-mercatoethanol 50 mM ;
- Solution to measure the sulphatase activity : SINIGRIN 0,15 mM in a buffer solution pH 5,8 prepared from ethylene diamine and acetic acid solutions 33 mM.

3. PREPARATION OF THE ENZYMATIC SOLUTION

It is preferable to purify the commercial sulphatase to increase its specific activity and avoid the clogging of ion-exchange resins DEAE Sephadex A25.

Two methods of purification A and B are used. They were described previously by THIES (7. THIES.1980) and HEANEY and FENWICK (8. HEANEY et al. 1980).

A - Method by precipitation/centrifuging and ion-exchange chromatography used in Canada and Britain

B - Method by ion-exchange chromatography on DEAE Sepharose Cℓ-6B and filtering used in Western-Germany and France.

The activity unit U was defined as follows to compare both methods : 1 U desulphate 1 µmol of SINIGRIN per minute at 30°C and pH 5,8.

3.1. MEASURE OF ACTIVITY

The wave length of SINIGRIN maximum absorption recorded on our spectrophotometer was : 228nm.

Therefore, activity will be calculated by measuring the decrease in sinigrin absorption at this wave length.

Working method :

Transfer 2ml of solution to the 2 spectrophotometer-cuvettes to measure sulphatase activity. At the time t=0, transfer 50 µl of the purified sulphatase solution to the measure cuvette.

Trace the curve A = f(t), using if possible a recorder.

The obtained curve looks like below :

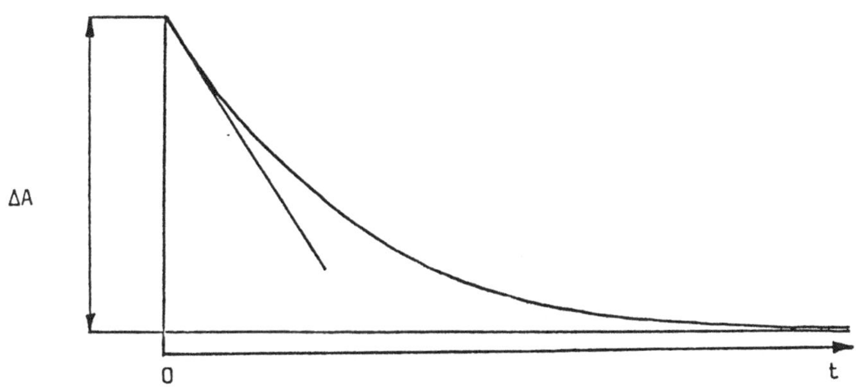

ΔA is the difference of absorption due to the transformation of SINIGRIN into DS SIN. If c is the concentration of DSIN and Δε ,the difference of molecular extinction coefficients of SIN and DSIN

$$\Delta A = \Delta\varepsilon . \ell . c .$$

When the reaction-balance is reached, C represents 95% of the SIN initial concentration (4. THIES 1978). This value was checked by an HPLC-analysis of the desulphated solution.

$$C = \frac{0,15 \times 10\text{-}3 \times 0,95 \times 2}{2,05} = 0,139\text{mM}$$

When the reaction-balance is reached, $\Delta A = 0,1925$.

$$\Delta\varepsilon = \frac{0,1925}{0,139 \times 10\text{-}3 \times \ell} = 1385 \ \ell.\text{mol-1 cm-1}$$

The DSIN concentration can be calculated at any time in the desulphation process :

$$C = \frac{\Delta A}{1385 \times \ell} = \frac{\Delta A}{1385} \quad (\ell = 1 \text{ cm})$$

During Δt, the quantity of desulphated SIN contained in the cuvette is equal to : C x V (V : volume of the reaction mixture in l) :

$$C \times V = \frac{\Delta A.V}{1385}$$ with ΔA : observed decrease during Δt (Δt in mn)

$$\text{Activity : } \frac{\Delta A.v}{1385.\ t} = \frac{V}{1385} \times \frac{\Delta A}{\Delta t}$$

where $\frac{\Delta A}{\Delta t}$ = tangent slope at the point t = 0.

3.2. RESULTS OF THE SULPHATASE ACTIVITY MEASURE PREPARED BY BOTH METHODS A AND B.

	Sulphatase A	Sulphatase B
Concentration	3,5%	1%
Activity	0,035 U	0,030 U
Activity/ml	0,70 U/ml	0,60 U/ml
Specific activity	0,02 U/mg	0,06 U/mg

These results led us to conclude that sulphatase prepared by the Method B (purification by DEAE Sepharose and filter Millipore 10000 daltons) allows us to obtain a specific activity which is three times higher. That is why we shall use this method for the rest of our study.

4. OPTIMAL CONDITIONS OF DESULPHATION

4.1. pH-INFLUENCE.

Measures are made in the spectrophotometer-vats as said before.

pH	4,5	5,0	5,4	6,0	6,5	7,0	7,5
Activity (in U)	0,038	0,043	0,044	0,038	0,022	0,011	0,003

Table 1.

The used buffer is diamine ethylene/acetic acid. The optimum pH corresponds to the value recorded by THIES (3. THIES 1979). We have to note the non-measurable activity at pH 8,0.

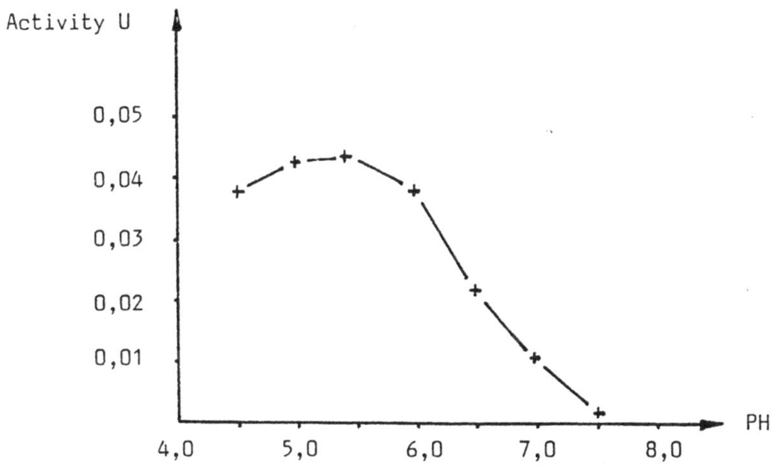

Fig.1.

4.2. EFFECT OF TEMPERATURES

The enzymatic solution is pre-warmed at the studied temperature for 10 minutes to suppress the inertia effect of enzyme in its reaction to heat.

T°C	25	29	33,5	38	42	46,5	50
Activity	0,042	0,058	0,070	0,093	0,091	0,087	0,096
	60	65	70				
	0,096	0,017	≈ 0				

Table 2.

Sulphatase has a maximum activity at 50°C. This high value may be explained by the purification process and the presence of a buffer. It is not certain that this resistance will still exist on ion-exchange columns where impurities and ions are present. Therefore, it will be advisable not to go beyond 45°C in analyses to come. The choice of this temperature will also be guided by the stability of glucosinolates and their desulphated derivatives.

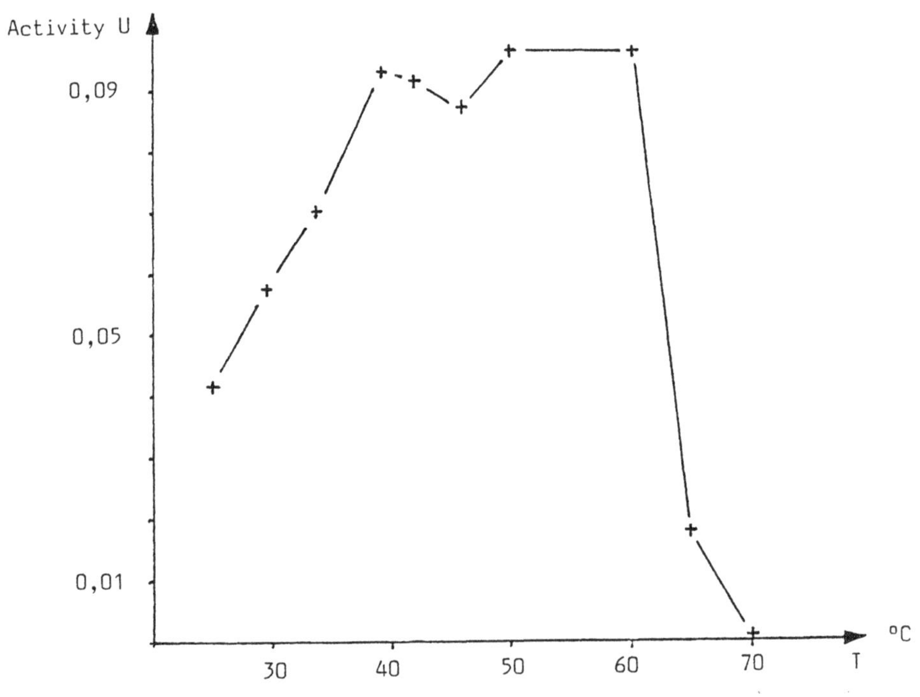

Fig.2.

5. DESULPHATION ON ION-EXCHANGE COLUMNS.

5.1. KINETICS.

This study was carried out on a 00-winter rapeseed sample. The extraction of glucosinolates was performed following Sang and Truscott (6. SANG et al. 1984) in the presence of a complexing agent EDTA and an anti-oxidising agent-2-mercapto-ethanol, the latter leading to a reduction of 4-OH-glucobrassicin-losses.

Working method :

After preheating at 95°C for 2 minutes, 200 mg of crushed seeds were extracted at 90°C in a 5ml-polypropylene tube by 1,5 ml of solution EDTA 1mM, 2-mercaptoethanol 50 mM.

After stirring, 500µl of sinigrin 2 mM were added. The whole extraction lasted 5 minutes.

After a first centrifugation at 5000 revolutions per minute for 5 minutes, 1 ml of supernatent, 0,25 ml of lead and barium acetate 0,5 M, was added. It was followed by another centrifugation at 5000 revolutions per minute for 5 minutes. 1 ml of supernatant was then taken out and diluted four times with water before transferring it to the columns of DEAE Sephadex.

The columns were washed with 1 ml of the desulphation solution A at pH 5,8 and 250 µl of sulphatase at 0,1% in this same solution were added. The columns were then choked and placed at 37°C.

After a definite reaction-time, glucosinolates were eluted by 3ml of water. The DSGLS-solution was ready for injection.

DESCRIPTION OF ION-EXCHANGE COLUMNS.

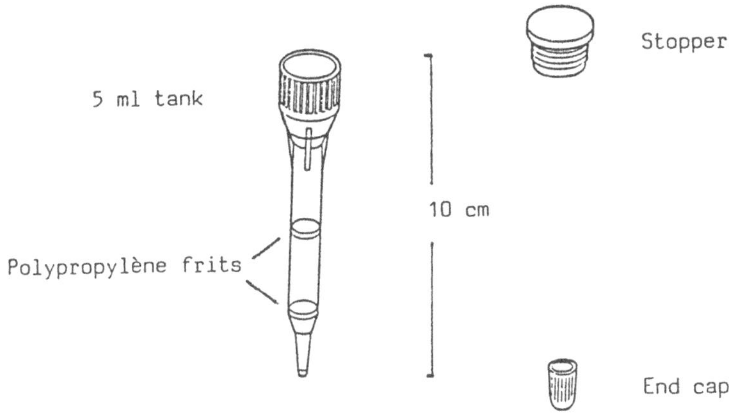

A dry weight of 50mg of DEAE Sephadex offers a height of 12 mm.

CHROMATOGRAPHIC CONDITIONS.

Column 250 x 4 mm ODS 2 5µ Spherisorb

Pre-column 30 x 4 mm ODS 5µ Spherisorb

Flow 1,5 ml/mn

Elution gradient Solvent A : water
Solvent B : Acetonitrile 25%

for 2 minutes A : 85%
B : 15%

from 2 to 18 min linear gradient to A : 0%
B : 100%

from 18 to 23 min A : 0%
B : 99%

from 23 to 28 min linear gradient to A : 85%
B : 15%

from 28 to 33 min A : 85%
B : 15%

UV detection at 229nm

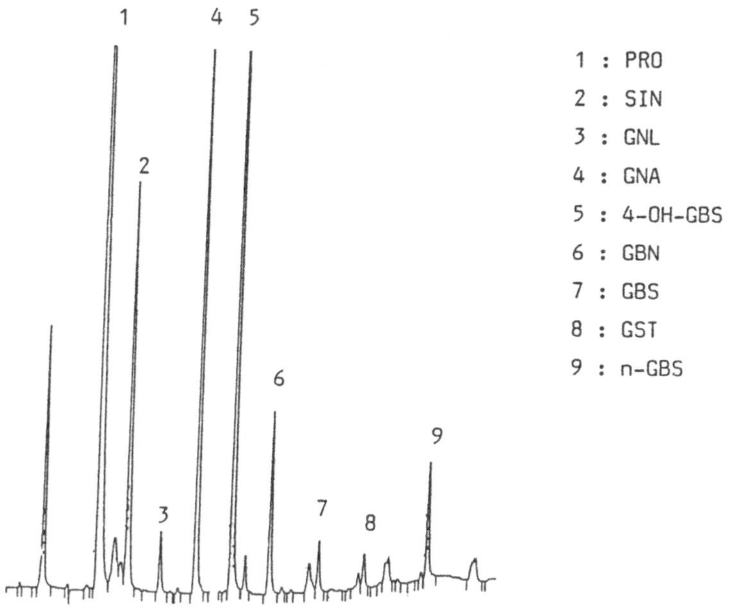

Fig.4

RESULTS

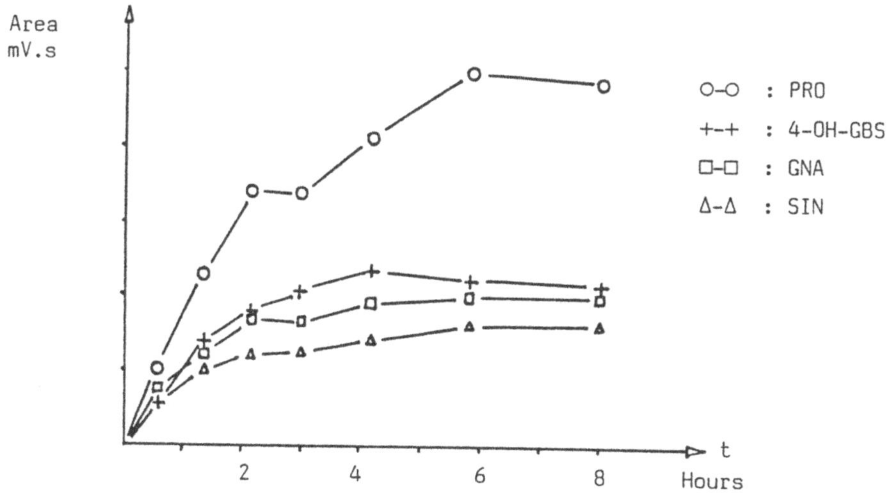

Fig.5

We notice that the necessary period of time to obtain maximum areas is about 6 hours, except for 4-OH-glucobrassicin for which 4 hours are enough. Then the area of this peak decreases appreciably because of the degradation of this compound. It is therefore probable that the maximum observed does not correspond to the real quantity of 4-OH-glucobrassicin.

5.2. COMPARISON OF SEVERAL METHODS.

Several methods of desulphation used in different labs are applied to the same extract prepared as said above.

DESULPHATION PROTOCOLS

Sample A	Sample B	Sulphatase	Concentration	Washing Solution	T°C
1	6	Crude	0,2% in buffer B	1 ml buffer B	37°C
2	7	purified B	0,1% in buffer B	1 ml buffer B	37°C
3	8	purified B	0,1% in buffer A	1 ml buffer A	37°C
4	9	purified B	0,1% in buffer A	1 ml buffer A	25°C
5	10	purified A	3,5% in water	2 ml water + 1 ml buffer C	25°C

Table 3

Purified B : method described by W. THIES

Purified A : method described by Mc GREGOR/FENWICK

Buffer A : EDA/CH_3COOH 20mM pH 5,8
EDTA 0,1 mM
2-mercaptoethanol 5 mM

Buffer B : TRIS 18 mM pH 8,0
EDTA 0,1 mM
2-mercaptoethanol 5 mM

Buffer C : pyridine acetate 0,02 M

Sample A : ≃50 µMol/g seed

Sample B : ≃25 µMol/g seed

RESULTS

1) ALKENYLS GLS (PRO, GNA) □ PRO ▨ GNA

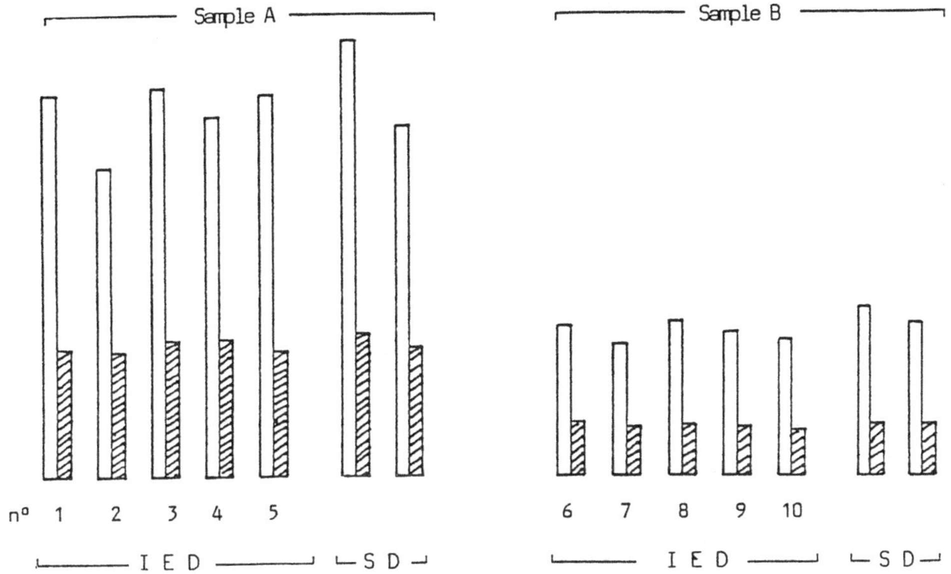

2) INDOLYLS GLS (4-OH GBS, Neo GBS) □ 4-OH GBS ▨ néo GBS

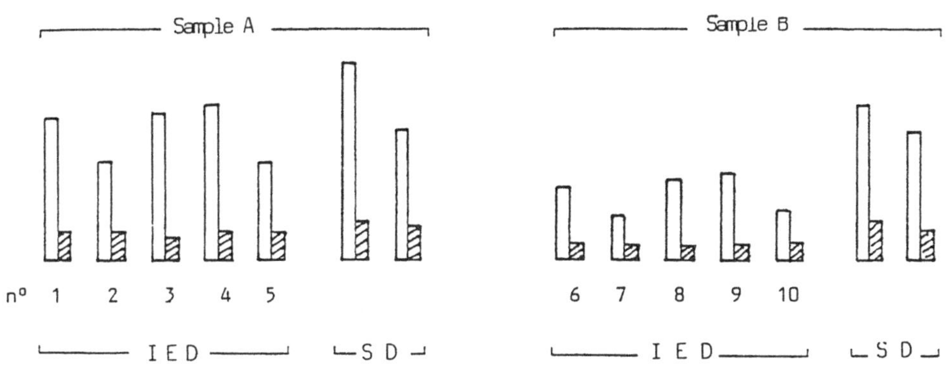

I E D : Ion Exchange Desulphation
S D : Solution Desulphation

Fig.6.

It has to be noted that :

- the profiles are fairly reproducible from the sample A to the sample B with quite different contents ;
- the use of an anti-oxidising agent is advisable for desulphation, for the content of 4-OH-GBS is slightly increased ;
- desulphations on non-purified extracts give the best results, especially for 4-OH-GBS. These desulphations were carried out as described below, and the extraction, carried out as said before. The raw extract is not diluted and has to be analysed quickly. It is filtered at 0,22 µ.

Desulphation : 50 µl/extract are added to 50 µl/sulphatase 1% After incubating at 25°C for 2 min., 20 µl are injected for the analysis.

In spite of the presence of impurities, the quantitative analysis of the main glucosinolates does not raise special problems. GBS alone, which is contained in small quantities in rapeseed seeds, cannot be quantified. Peaks appear when sulphatase is added. They correspond to desulphated glucosinolates.

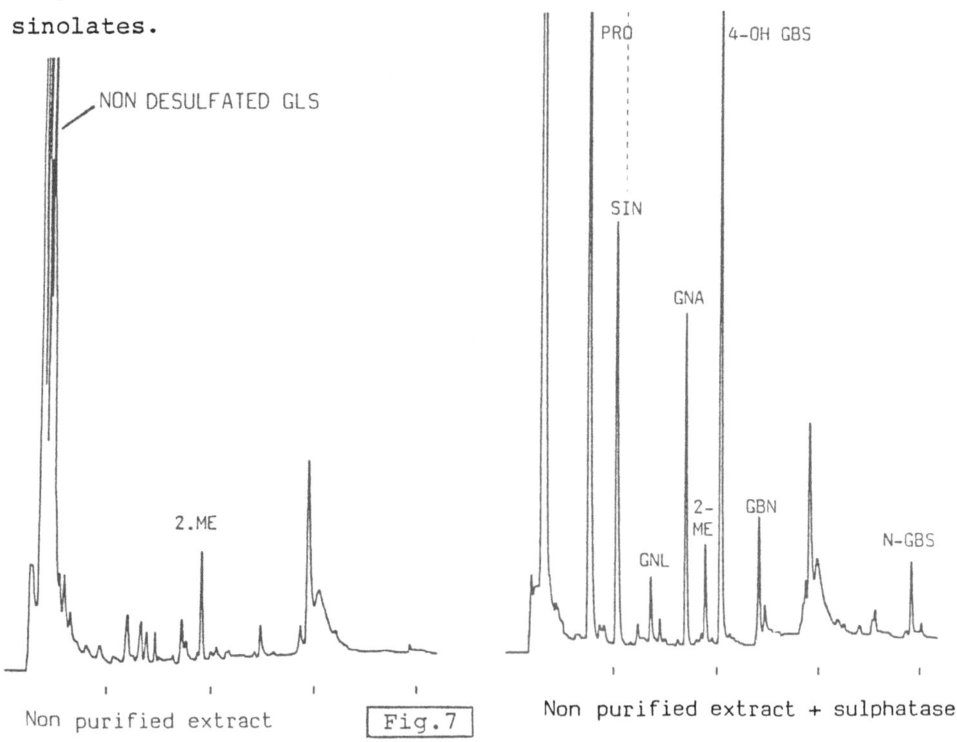

Fig.7

The best yields in DSGLS observed in the case of raw extracts may be explained :

- by a desulphation speed, which is greater in solution than on ion-exchanger ;
- by the fact that indolglucosinolates are not exposed to degradation risks very long (max.30 min. instead of several hours on ion-exchange columns).

Therefore, this technique is quite interesting for it gives us the possibility to approach the real value of 4-OH-GBS. That is why it is desirable to study desulphation kinetics on raw extracts.

5.3. DESULPHATION KINETICS OF NON PURIFIED GLUCOSINOLATE EXTRACTS.

The protocol was described previously. Desulphation time varied between 30 sec. and 4 min.

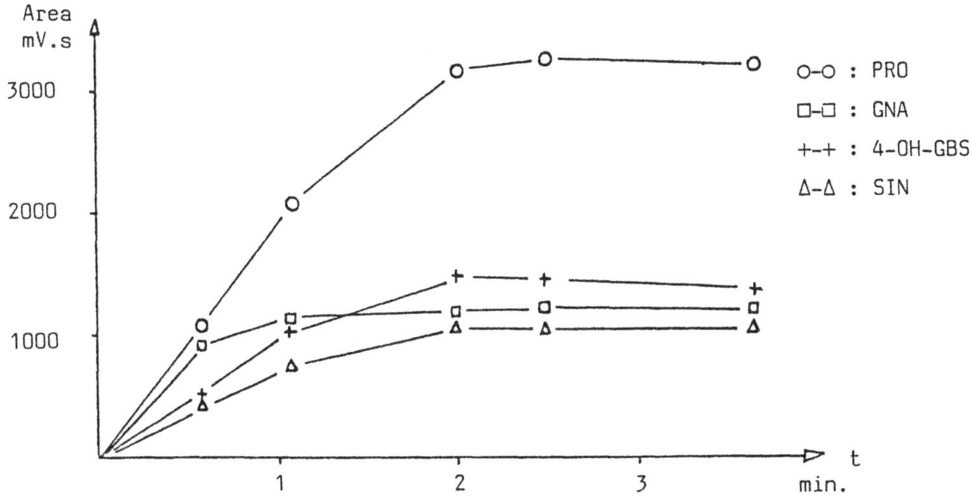

Fig.8

The optimal desulphation time is 2 min. The yields in DSGLS are higher than yields on ion-exchange columns.

Therefore, this method leads to a better quantitative analysis of glucosinolates. Present impurities are not a nuisance if the extract is filtered and the chromatographic system provided with a guard pre-column.Thus, it is possible to perform a great number of analyses without contaminating the analytical column. The duration of analysis is considerably reduced since the result can be obtained within 1 hour. On the other hand, the simultaneous extraction of several samples is not recommended, for the filtered raw extract must be desulphated and chromatographed in the following minutes. Moreover, the peaks of certain glucosinolates are lost in the peaks of impurities. Thus, glucobrassicin is not measurable. In certain cases, it will be better to carry out a "blank" analysis with the non desulphated extract to identify a peak.

This method has therefore its limits, but it can complement the method by desulphation on ion-exchangers. It is a fact that it gives better results for most glucosinolates.

6. INFLUENCE OF PRECIPITATION BY LEAD AND BARIUM ACETATE.

The extract treatment by these salts creates important precipitates of proteins, organic acids and sulphates detrimental to the enzymatic activity (4. THIES. 1978. and 7. THIES. 1980).

Extracts treated or untreated by barium acetate, lead acetate, und lead and barium acetate, were studied :

- after desulphation on ion-exchanger ;
- after desulphation in solution.

6.1. DESULPHATION ON ION-EXCHANGER.

Precipitation with	PRO	SIN	GNA	4-OH-GBS	GBN	N-GBS
untreated	2058	793	791	582	270	133
Ba-acetate	1729	696	714	586	246	118
Pb-acetate	2332	864	790	628	267	119
Ba- and Pb-acetate	2187	855	807	741	271	124

Table 4.

Results expressed in integration unit of peak areas.

6.2. DESULPHATION OF A RAW EXTRACT IN SOLUTION.

Duration of desulphation : 1 min.

Precipitation with	PRO	SIN	GNA	4-OH-GBS	GBN	N-GBS
untreated	351	351	412	NM	NM	66
Ba-acetate	873	292	581	551	231	NM
Pb-acetate	922	310	650	573	264	132
Ba- and Pb-acetate	1315	444	720	878	278	115

NM = not measurable

Table 5

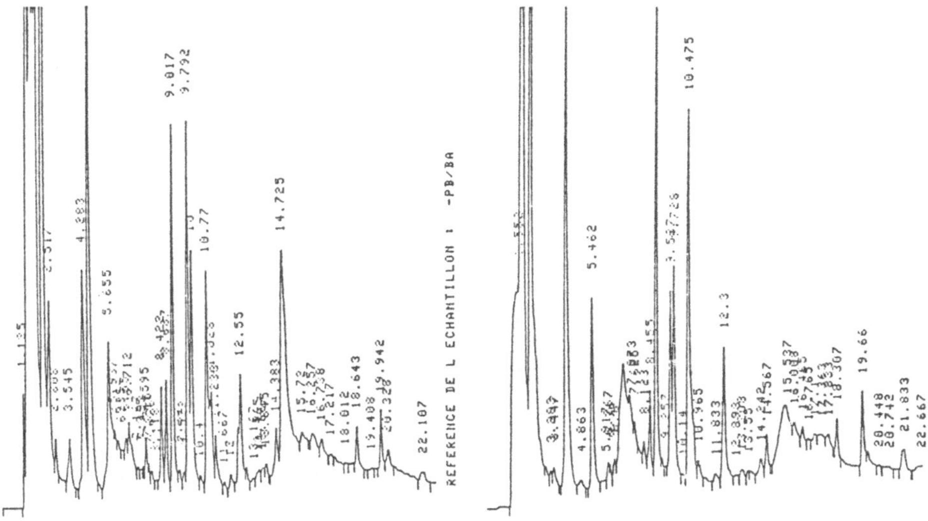

Untreated extract

Extract treated by barium acetate

Fig.9

Fig.9 (continued)

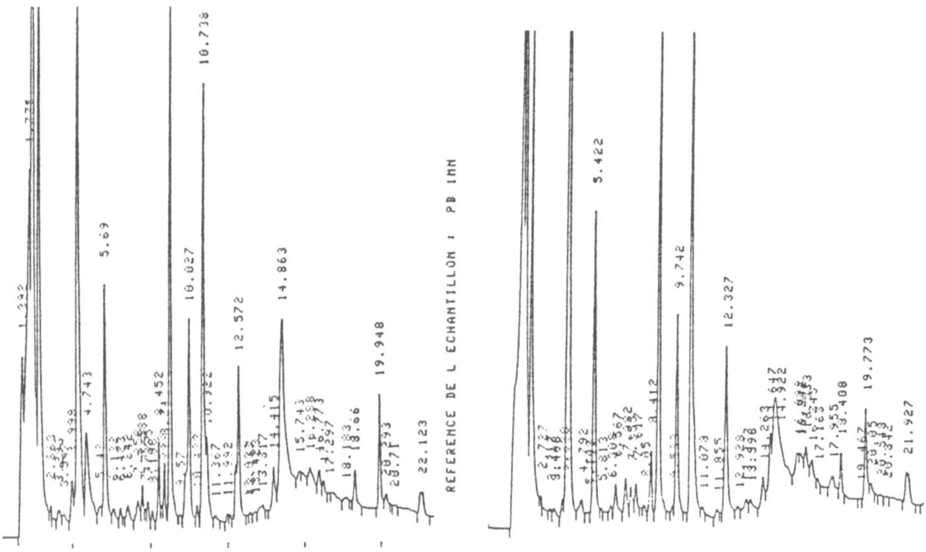

Extract treated by lead acetate

Extract treated by lead and barium acetate

DISCUSSION

The addition of lead acetate, barium acetate, or lead and barium acetate, has a positive effect on desulphation speed of raw extracts. It has less effect on a desulphation on ion-exchanger, for washing carries away a great amount of inhibitory substances. Precipitation also causes the disappearance of peaks of impurities, certain of which may be awkward on the chromatogram.

7. CONCLUSION

Glucosinolate desulphation on ion-exchangers is an elegant way of purifying and derivating glucosinolates. Unfortunately, the duration necessary for this process makes the quantitative analysis of 4-OH-glucobrassicin uncertain. A desulphation of raw extracts, which were only purified by precipitation with lead and barium acetate, gives us an interesting quantitative approach. 4-OH-glucobrassicin being integrally desulphated quite rapidly, the real content of this glucosinolate will be defined. Then, it will be possible to optimize desulphation on ion-exchange column, using an acid pH as well as an antioxidant (2-mercaptoethanol) and an adequate incubation temperature. Such conditions will also be interesting for the other analytical steps, such as extraction and chromatography, where the stability of 4-OH-glucobrassicin is insufficient.

BIBLIOGRAPHY

1. MOLLER P., OLSEN O., PLOGER A., RASMUSSEN K.W., SORENSEN H. (1984)
Quantitative Analysis of Individual Glucosinolates in Double Low Oilseed Rape by HPLC of Intact Glucosinolates, in : Advances in the Production and Utilization in Cruciferous Crops. pp. 111-126 (Proceedings of a Seminar in the CEC Programme of Research on Plant Protein Improvement. Held in Copenhagen, 11-13 September 1984). Martinus Nijhoff/Dr.W. Junk Publishers.

2. SPINKS A., SONES K., FENWICK G.R. (1984)
The Quantitative Analysis of Glucosinolates in Cruciferous Vegetables, Oilseeds and Forage Crops Using High Performance Liquid Chromatography.
in : Fette, Seifen, Anstrichmittel. 86, 228-231

3. THIES W. (1979)
Detection and Utilization of a Glucosinolate Sulfohydrolase in the Edible Snail, Helix Pomatia.
in : Naturwissenschaften, 66, 364-365

4. THIES W. (1978)
Quantitative Analysis of Glucosinolates after their enzymatic Desulphation on Ion Exchange Column.
in : Proc. 5th Int. Rapeseed Conference. Malmö, Sweden, June 12-16, 1978. 1, 136-139.

5. QUINSAC A. , RIBAILLIER D. (1984)
Quantitative Analysis of Glucosinolates in Rapeseed Seeds. Optimization of Desulphation.
in : Advances in the Production and Utilization of Cruciferous Crops. pp.85-110. Proceedings of a Seminar in the CEC Program of Research on Plant Protein Improvement. Held in Copenhagen, 11-13 September 1984). Martinus Nijhoff/Dr. W.Junk Publishers.

6. SANG. J.P., TRUSCOTT R.J.W. (1984)
Liquid Chromatographic Determination in Rapeseed as Desulfo-glucosinolates.
in : J. Assoc. Off. Anal. Chem. (Vol. 67, 829-833)

7. THIES W. (1980)
Analysis of Glucosinolates via on Column Desulfation, .
in : Proc. of a Symposium on Analytical Chemistry of Rapeseed and its Products. Held in Winnipeg, Canada, May 5-6,1980.

8. HEANEY R.K., FENWICK G.R. (1980)
The Analysis of Glucosinolates in Brassica Species Using Gas Chromatography. Direct Determination of the Thiocyanate Ion Precursors, Glucobrassicin and Neoglucobrassicin.
in J. Sci. Food. Agric. 31, 593-599

ANALYSIS OF INDIVIDUAL GLUCOSINOLATES IN RAPESEEDS COMPARISON BETWEEN DIFFERENT METHODS

J-P. Wathelet[+], R. Biston[++], M. Marlier[+] and M. Séverin[+]

\+ Faculté des Sciences Agronomiques de l'Etat
Chaire de Chimie Générale et Organique
B 5800 Gembloux - Belgium

++ Centre de Recherches Agronomiques de l'Etat
Station de Haute Belgique
rue de Serpont, 48
B 6600 Libramont - Belgium

ABSTRACT

The aim of our work is to contribute to the elaboration of a reliable quantitative method for individual glucosinolates determination in rapeseeds.

To emphasize the accuracy of the method, we have studied several important points :

a) methanol - water (70%) extraction from seeds and seedcakes obtained after removal of the oil.

b) column chromatographic purification of crude extracts

- the influence of the height of D.E.A.E. A 25 columns (desulphoglucosinolates) and of Ecteolla Cellulose (intact glucosinolates) on the level of glucosinolates obtained is analysed
- to avoid the use of pyridin, the pyridin - acetate buffer frequently used to prepare D.E.A.E. columns is changed by an acetic acid - sodium acetate buffer (pH : 5)

c) HPLC analysis of desulphoglucosinolates and intact glucosinolates

- comparison of desulphoglucosinolates and intact glucosinolates separations
- choice of the detection wavelength
- influence of column temperature

d) Identification of desulphoglucosinolates from different cultivars of rapeseed (Garant, Ledos, Lirasol and Librador).

INTRODUCTION

Glucosinolates are easily hydrolysed by the enzyme myrosinase to yield an unstable aglucone which breaks down to give a range of products such as nitriles, isothiocyanates, oxazolidine thiones and so on.

To valorize the rapeseed in the best way, it is imperative to reduce the level of glucosinolates by breeding new varieties. It is also important to use precise and reliable analytical methods.

Although the official method is the measurement of glucosinolates by GLC with temperature programmation, Spinks, Sones and Fenwick (Spinks at al., 1983) , Helboe, Olsen and Sørensen (Helboe et al., 1980) and Møller , Olsen, Plöger, Rasmussen and Sørensen(Møller et al., 1986) have developped two different methods by HPLC. The first one is separating the desulphoglucosinolates, whereas the second one measures directly the intact glucosinolates.

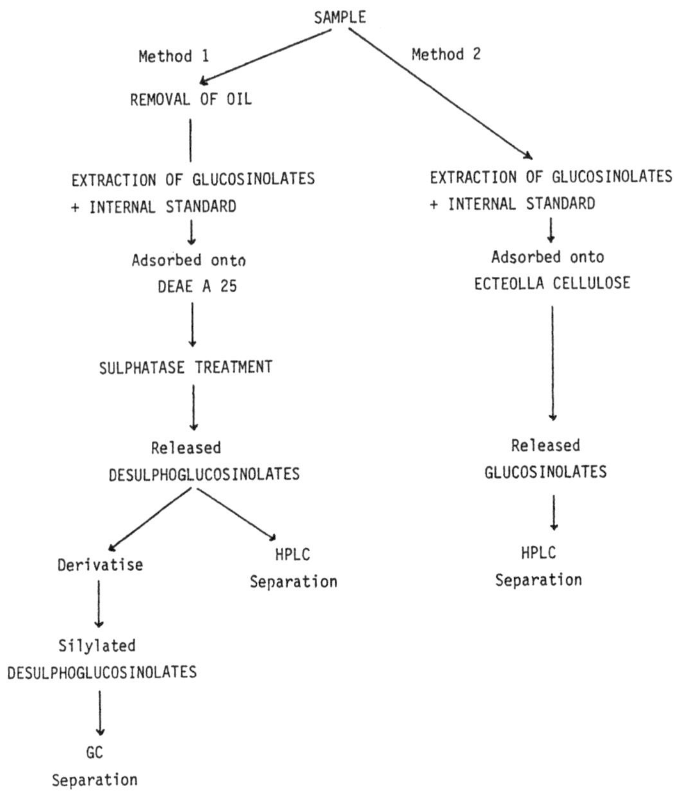

The purpose of this lecture is to describe and to compare the two HPLC methods we have tested in our laboratory.

RESULTS

Extraction - Removal of the oil

The extraction according to Sørensen method is really faster and avoids the use of expensive solvents to remove the oil. The question is whether it is necessary to remove the oil.

In this way, four rapeseeds containing different levels of glucosinolates have been analysed with or without removing the oil. Glucosinolates contained in the meal or in the seeds have been extracted with hot methanol water (70 %) .

TABLE 1 Extraction with or without removal of the oil

	A		B		C		D	
	DF	N.D	DF	N.D	DF	N.D	DF	N.D
PROGOITRIN	1.8	2.0	11.0	12.0	18.3	19.5	44.3	45.0
GLUCOBRASSICANAPIN	-	-	0.2	0.2	0.5	0.5	2.3	2.2
GLUCONAPIN	1.3	1.4	3.7	4.1	9.0	9.4	12.9	13.2
4 OH GLUCOBRASSICIN	4.5	4.6	3.7	3.6	3.6	3.6	3.8	3.8
GLUCONAPOLEIFERIN	-	-	0.7	0.7	0.5	0.5	2.0	2.0
GLUCOBRASSICIN	0.8	0.8	0.4	0.4	0.5	0.5	0.7	0.7

DF : with removal of oil

N.D.: without removal of oil

In a common way, the results are relatively close to oneanother (table 1). The percentage of 4 OH glucobrassicin doesn't really change but the concentration in progoitrin and gluconapin increases slightly when the extraction is being made without delipidation.

On the other hand, with thin layer chromatographies, we have demonstrated that the lipids remained absorbed on the D.E.A.E. A.25 and Ecteolla columns.So these fats are not able to pollute the HPLC columns. According to these results, we are proposing to follow the way without delipidation to extract the glucosinolates.

Purification columns

When the extraction is over, two different ways are used; either to desulphate the glucosinolates on the D.E.A.E. sephadex A 25 or to purify directly the intact glucosinolates on Ecteolla cellulose columns.

D.E.A.E. columns : concerning the first method, glucosinolates are absorbed on D.E.A.E. sephadex columns prepared using pasteur pipettes with a pyridin acetate buffer.

In our laboratory, we have realized experiments to determine if the quantity of resin set in the column is influencing the results.

Four rapeseeds have been extracted. Columns containing 15, 25 and 35 mg of sephadex resin have been tested (Table 2).As table 2 shows, the results are not really different.

Pyridin used in the buffer is a highly toxic product which, following the commercial origin, can give a very large peak after 17 minutes of analysis by HPLC. This interferes considerably with the measurement of gluconasturtiin and glucobrassicin.

TABLE 2 Influence of the height of D.E.A.E. sephadex column.

	A			B			C			D		
MG of D.E.A.E	15	25	35	15	25	35	15	25	35	15	25	35
PROGOITRIN	1.7	1.7	1.8	12.2	12.2	12.3	18.5	17.6	17.7	44.9	44.3	44.0
GLUCOBRASSICANAPIN	-	-	-	0.2	0.2	0.2	0.5	0.5	0.5	2.3	2.3	2.3
GLUCONAPIN	1.2	1.2	1.2	4.3	4.3	4.3	9.0	8.5	8.5	13.0	12.9	12.9
4 OH GLUCOBRASSICIN	5.2	5.5	5.0	4.2	4.3	3.7	3.8	3.6	3.6	3.8	3.7	4.0
GLUCONAPOLEIFERIN	-	-	-	0.7	0.6	0.6	0.5	0.5	0.5	2.0	2.0	1.9
GLUCOBRASSICIN	0.8	0.8	0.8	0.4	0.4	0.4	0.6	0.6	0.6	0.7	0.7	0.8

So, in collaboration with Professor Sørensen, we have conducted research to find a new buffer.

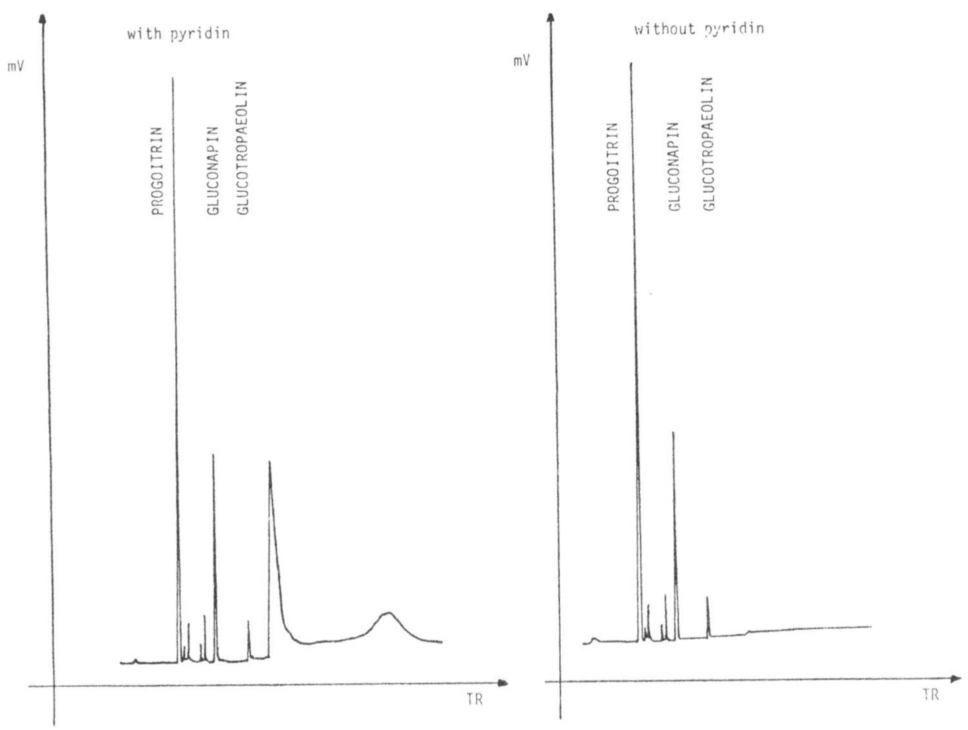

Fig. 1 Comparison of buffers used to prepare D.E.A.E. columns

The replacement of pyridin acetate buffer by acetic acid - sodium acetate buffer at pH 5 allow's us to overcome this obstacle (Fig. 1).Table 3 shows that generally both buffers give similar results by HPLC.

TABLE 3 Comparison of buffers used to prepare D.E.A.E. columns

	A		B		C		D	
	Pyr	HAc	Pyr	HAc	Pyr	HAc	Pyr	HAc
GLUCOIBERIN	3.8 (glucoiberin + glucoraphanin)	1.1	2.5	2.5	0.4	0.4	16.5	13.4
GLUCORAPHANIN		2.5	2.0	2.0	0.3	0.4	30.8 (glucoraphanin + sinigrin)	10.7
SINIGRIN	-	-	-	-	-	-		18.8
GLUCOALYSSIN	2.8	3.0	1.9	1.9	1.4	1.4	-	-
PROGOITRIN	45.6	46.3	89.6+	92.7+	14.5	13.7	11.5	12.4
GLUCONAPOLEIFERIN	1.3	1.1	5.8	5.8	0.8	0.8	0.6	0.9
GLUCOIBERVIRIN	-	-	-	-	-	-	3.9	3.9
GLUCONAPIN	14.9	14.7	25.5	26.5	5.9	5.9	2.4	2.7
GLUCOERUCIN	-	-	-	-	-	-	6.3	6.5
SINALBIN	-	-	-	-	6.2	6.3	-	-
4 OH GLUCOBRASSICIN	-	-	7.1	7.9	5.8	6.2	4.6	4.6
GLUCOBRASSICANAPIN	3.5	3.3	7.4	7.3	1.9	1.9	-	-
NEOGLUCOBRASSICIN	-	-	3.9	2.1	-	-	-	-
GLUCOBRASSICIN	-	-	-	0.2	-	1.4	-	4.0

Pyr : pyridin buffer
HAc : acetic acid buffer
\+ : unknown + progoitrin

<u>Ecteolla columns</u> : according to the second method, the intact glucosinolates are absorbed in Ecteolla cellulose columns.

The height of the columns as well as the quantity of sodium hydrogencarbonate used to eluate glucosinolates are very important. A small column doesn't keep enough of the alkenyl glucosinolates, a too high column doesn't allow the complete elution of the 4 OH glucobrassicin. Therefore it is necessary to find a compromise.

DESULPHATION

The action of Helix Pomatia sulphatase is an extra step that allows the transformation of glucosinolates in to desulphoglucosinolates. The preparation of the enzyme and the desulphation must be well controlled if we want to obtain good reproducibility. In our laboratory, sulphatase is purified following the method described by Heaney (Heaney et al.,1986).

GAS CHROMATOGRAPHY - GC-MS

By using the first method (desulphoglucosinolates), it is possible to analyse desulphoglucosinolates with GLC but also by HPLC.

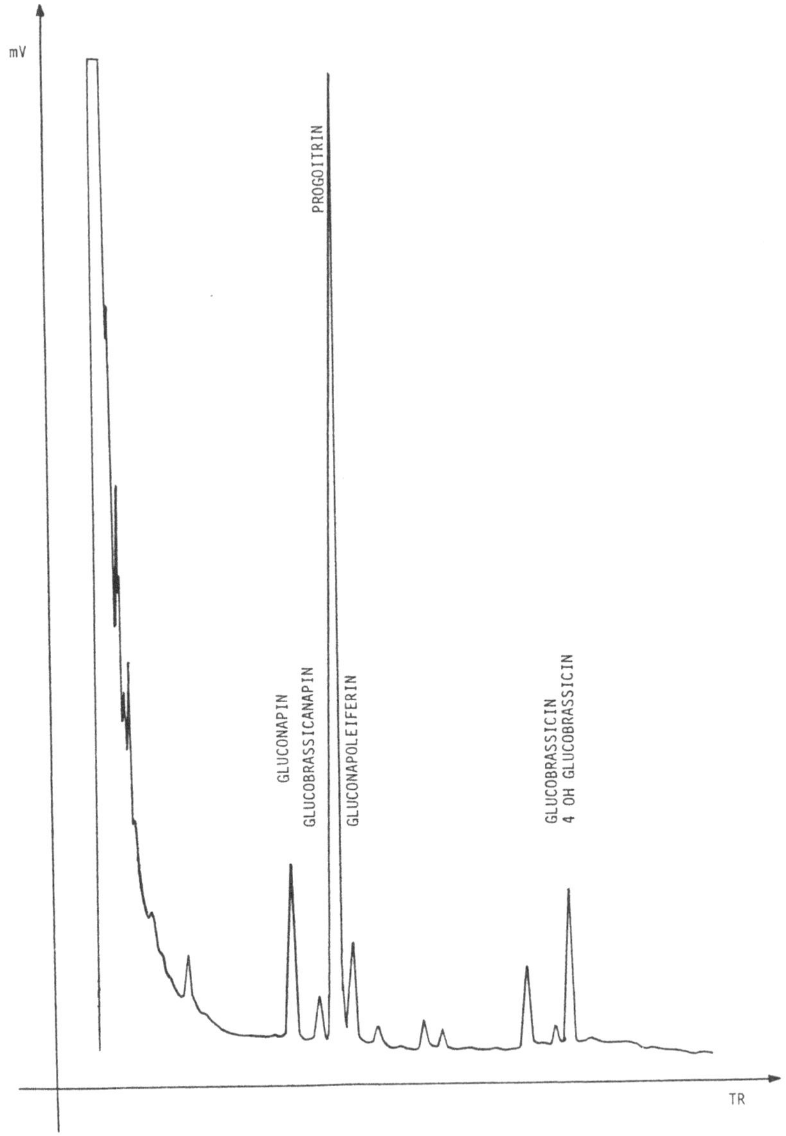

Fig. 2 Separation of silylated desulphoglucosinolates by GLC

Concerning GLC analysis, we are using as silylating reagent a mixture of MSHFBA, TMCS and 1 methylimidazole. With this reactive, silylation is quick. After 7 or 8 minutes at 120°C, the reaction is finished.

The silylated derivatives are separated in a column of OV 17 on diatomite CLQ. In these conditions, good separation of the silylated desulphoglucosinolates can be obtained(Fig.2).

Glucosinolates isolated from four rapeseed varieties (Garant, Lirasol, Librador and Ledos) have also been studied by GC-MS in electron impact and chemical ionisation. The goal is to confirm the identity of glucosinolates extracted from these seeds and to try to identify unknown peaks.

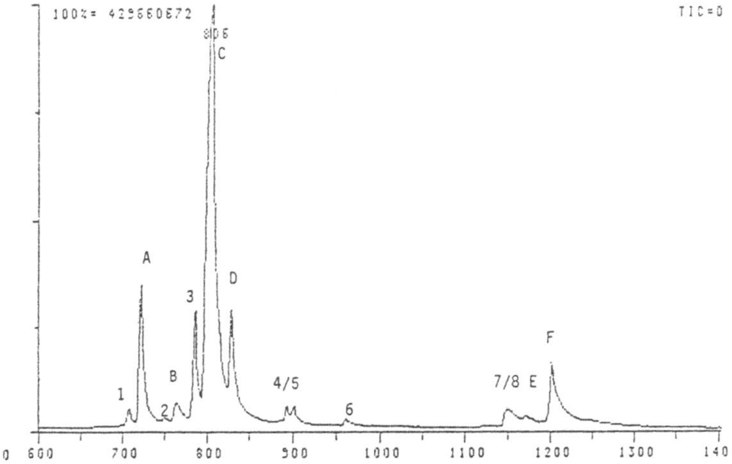

Fig. 3 Silylated desulphoglucosinolates from Ledos
Total ionic current (GC-MS)

The analytical conditions in GC-MS are the following : the column is a capillary CP SIL 5 of 20 meters. It is brought close to the source of the spectrometer through an open interface heated at 280°C. The temperature of the source of the spectrometer is 130°C and the ions energy is 70 ev. The reactant gas, in chemical ionisation is ammoniac under a pressure of 0.1 mm. The injection in the GLC is realized in splitless mode.

We have insisted on the detection of molecular ions. In this case, it is clear that the utilisation of chemical ionisation gives the best detection of these ions.

For the most common glucosinolates identified in the four rapeseeds (gluconapin, progoitrin, glucobrassicanapin, 4 OH glucobrassicin, gluconapoleiferin,glucobrassicin), we can see molecular ions in electron impact in optimum work conditions.Figure 4 shows the results obtained in electron impact conditions for gluconapin, glucobrassicanapin, progoitrin, gluconapoleiferin and 4 OH glucobrassicin.

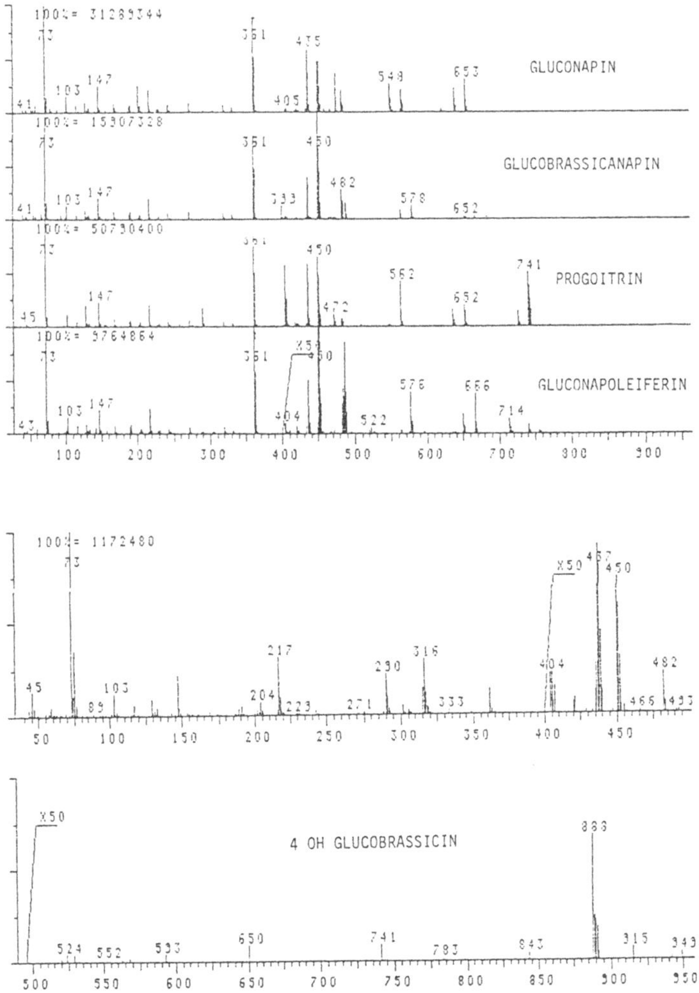

Fig.4 Mass spectra of silylated desulphoglucosinolates (electron impact)

Figure 5 shows the mass spectrum of progoitrin and gluconapoleiferin in chemical ionisation with ammoniac.

The characteristic fragments set up by Christensen (Christensen et al., 1982)for electron impact mass spectra are present in most of our spectra.We note anyway the extra presence of a M-41 peak in the silylated desulphogluconapoleiferin spectrum.

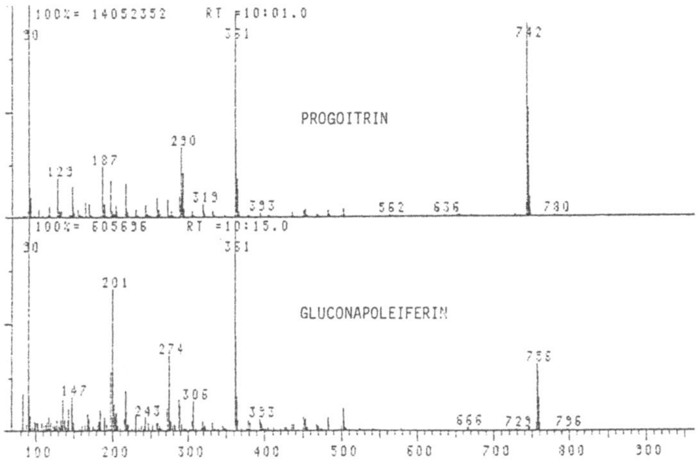

Fig. 5 Mass spectra of silylated desulphoglucosinolates (Chemical ionisation with ammoniac)

The chemical ionisation spectra always presents a big quasimolecular ion M+1. But we also find the characteristic mass peaks mentioned by Eagles (Eagles et al.,1981) in the same ionisation conditions.

In addition to the common glucosinolates labelled from A to F, we can see, on the chromatogram of Ledos variety (Fig.3), several more or less important peaks.

The compounds labelled 4,5,6,7 and 8 don't seem to be glucosinolates. Peaks 4,7, and 8 could be sugar or derivatives of sugar. Peak number 5 is visible in the blank; its mass is 388, it seems to be a TMS derivative. Peak number 6 is particular; molecular ion was not visible but it's electron impact spectrum gives an important ion with a 91 mass. That makes we think it is an aromatic derivative similar to glucotropaeolin.

Peaks 1,2 and 3 are situated just before the gluconapin, glucobrassicanapin and progoitrin respectively. Their molecular ions and their fragmentation is similar to those that follow although more intense (peaks A,B,C). We think that they could be isomers of gluconapin, glucobrassicanapin and progoitrin. We see them in different quantities in the four rapeseeds studied. We cannot eliminate the possibility of isomerisation in the whole analytical process (splitless injection, extraction, desulphation...).
In fact, the chromatograms of pure gluconapin and progoitrin extracted by Fenwick or Sørensen contain peaks having a similar retention time. But

we have not be able so far to determine the natural origin of these isomers.

HPLC of desulphoglucosinolates

HPLC separation : the desulphoglucosinolates obtained after action of the Helix Pomatia sulphatase are separated on a Spherisob ODS 2 column.The solvants used are water and acetonitrile at 20% as recommended by Heaney (Heaney et al.,1986).

In these conditions, we obtain in our laboratory very good separations and a very light variation of the baseline(Fig. 6)

Glucotropaeolin and glucobarbarin are convenient as internal standard. Nevertheless it would be desirable to find a company to commercialize them.

Moreover, we have realized very good reproducibility of retention times which is not to be neglected for a series of analyses with an integrator system and an automatic autosampler.

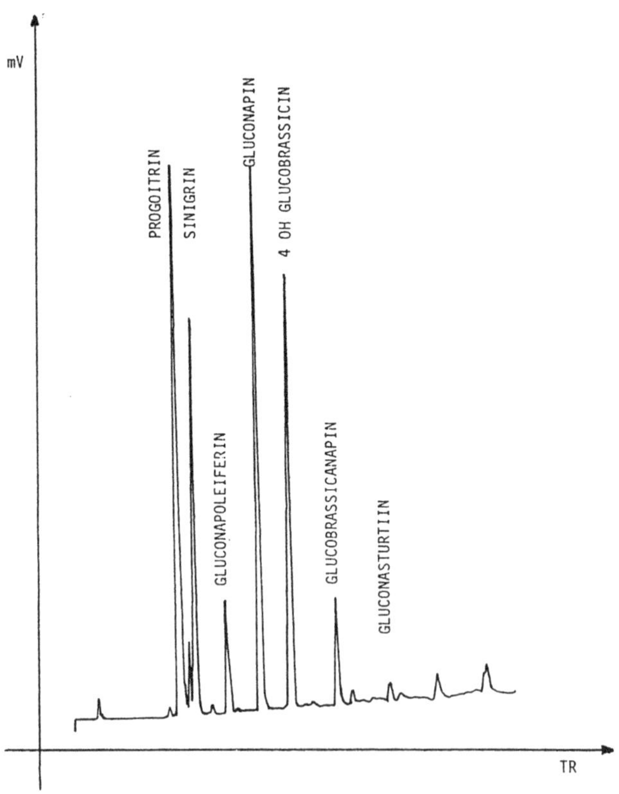

Fig.6 HPLC separation of desulphoglucosinolates.

Columns we are using in our laboratory have a long life; pressure column is also very stable.

u.v. spectra: u.v. spectra of pure desulphoglucosinolates or intact glucosinolates dissolved in water or in acetonitrile at 20% has been recorded with an u.v. spectrometer. On the other hand, our HPLC system is able to record u.v. spectra just after leaving the HPLC solumn. It is interesting to compare the results obtained (Table 4).

TABLE 4 Maximum wavelength observed for desulphoglucosinolates and intact glucosinolates.

	DESULPHOGLUCOSINOLATES			INTACTS	
	HPLC	WATER	AcCN	WATER	AcCN
GLUCOERUCIN	225				
GLUCOIBERIN		227	224	225	225
GLUCOCHEIROLIN	223	225	224	226	225
GLUCORAPHENIN		226	225	225	225
EPIPROGOITRIN	225	226	224	228	228
SINIGRIN	225	225	225	227	
GLUCONAPIN	223	224	223	226	226
GLUCOBARBARIN	230	229	227	227	228
PROGOITRIN	225				
2 RHAMNO PYRANOSYLOXY BENZYL GLUCOSINOLATE	217	218		215	215
SINALBIN	225	224	225	226	226
4 OH GLUCOBRASSICIN	221	222	220	221	222
GLUCOBRASSICIN	219	219	220	219	220
GLUCOTROPAEOLIN	232	229	230		
GLUCONASTURTIIN	232	223	230	226	227
NEOGLUCOBRASSICIN		221	220	218	218
m METHOXY GLUCOBRASSICIN	220	222	220	218	220
GLUCOCAPPARIN	221				

With these results, it seems that the best wavelength to use is between 225 and 230 nm. So we decided to work at 229 nm which is the wavelength fixed when using a cadmium lamp.

Influence of the HPLC oven temperature: At the beginning of our work, we have seen that the results for 4 OH glucobrassicin are not reproducible.

For the same solution values obtained for gluconapin, glucobrassicanapin, progoitrin and gluconapoleiferin were very close one to the other. We have realized that the HPLC oven temperature greatly influenced the measurement of 4 OH glucobrassicin.

Table 5 shows clearly that the concentration in 4 OH glucobrassicin measured at 230 nm with sinigrin as internal standard decreased in function of the temperature.

TABLE 5 Influence of the HPLC oven temperature

	A			B			C			D		
Temperature	25	30	35	25	30	35	25	30	35	25	30	35
PROGOITRIN	1.6	1.6	1.7	12.2	12.0	12.3	17.7	17.7	17.8	43.6	44.4	44.7
GLUCOBRASSICANAPIN	-	-	-	0.2	0.2	0.2	0.4	0.5	0.4	2.3	2.2	2.3
GLUCONAPIN	1.2	1.2	1.3	4.3	4.2	4.4	8.5	8.6	8.6	12.4	12.9	13.0
4 OH GLUCOBRASSICIN	5.4	4.4	3.3	4.6	3.6	2.8	4.3	3.4	2.5	4.2	3.4	2.6
GLUCONAPOLEIFERIN	-	-	-	0.6	0.7	0.6	0.5	0.5	0.5	1.9	2.0	2.0
GLUCOBRASSICIN	0.8	0.8	0.8	0.5	0.4	0.4	0.5	0.4	0.5	0.6	0.7	0.6

Is this loss in 4 OH glucobrassicin due to the degradation of the compound in the column? More complementary experiments should prove it. In any case it is important to fix the temperature of the HPLC column to obtain better reproducibility.

It is a routine for us now to use the desulphoglucosinolate method. A lot of samples coming from France, Denmark and Belgium have been analysed by GLC and HPLC with the aim of calibrating a near infrared reflectance spectrometer.

HPLC of intact glucosinolates

The intact glucosinolates eluated from Ecteolla columns are, without sulphatase treatment, directly injected in the HPLC in a Nucleosil 5 C18 column.

Methanol buffer as eluant: At the beginning of our experiments in this field, we were using as solvent a mixture of 0.01 M phosphate buffer at pH 7 containing 5 mM of tetraheptylammonium bromide and modified with 62.5 % methanol.

The isocratic system used requires only one pump; in this way we don't need very expensive equipment.

In these conditions, we have obtained more or less satisfactory separations with large and tailing peaks(Fig.7)

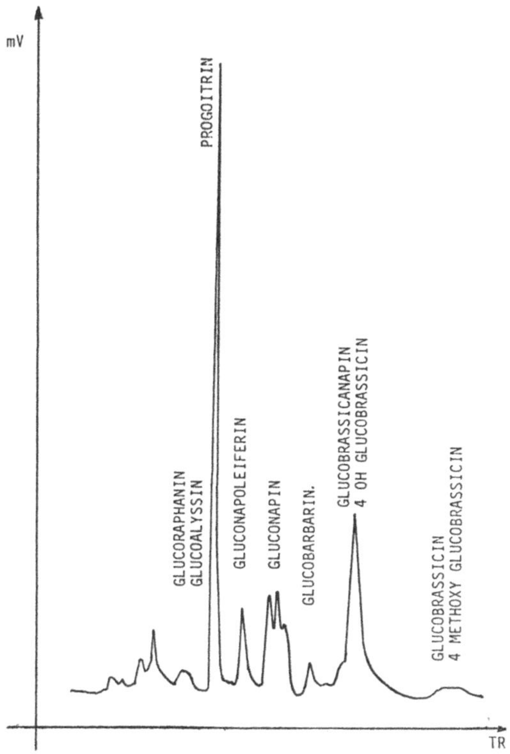

Fig.7 HPLC separation of intact glucosinolates
(Methanol buffer as eluant)

In this method, gluconapin is eluated at approximatively the same retention time as two other products which could be probably cinnamoyl derivatives of glucosinolates.We have seen these peaks in French (Tandem, Darmor) and Danish rapeseeds.

The glucobarbarin is convenient as internal standard but not commercialized.The glucotropaeolin can also be used as internal standard but glucobrassicanapin is eluated very close and it seems that in isolated cases, derivatives of glucosinolates are also interfering.

The use of a methanol buffer quickly increased the pressure of the

HPLC column. So the columns need to be changed very often which is expensive if we don't have all the necessary material to fill up those columns.

Acetonitrile buffer as eluant: Prof. Sørensen has recently switched the methanol buffer to an acetonitril one.

In these conditions the separation is well improved (Fig. 8)

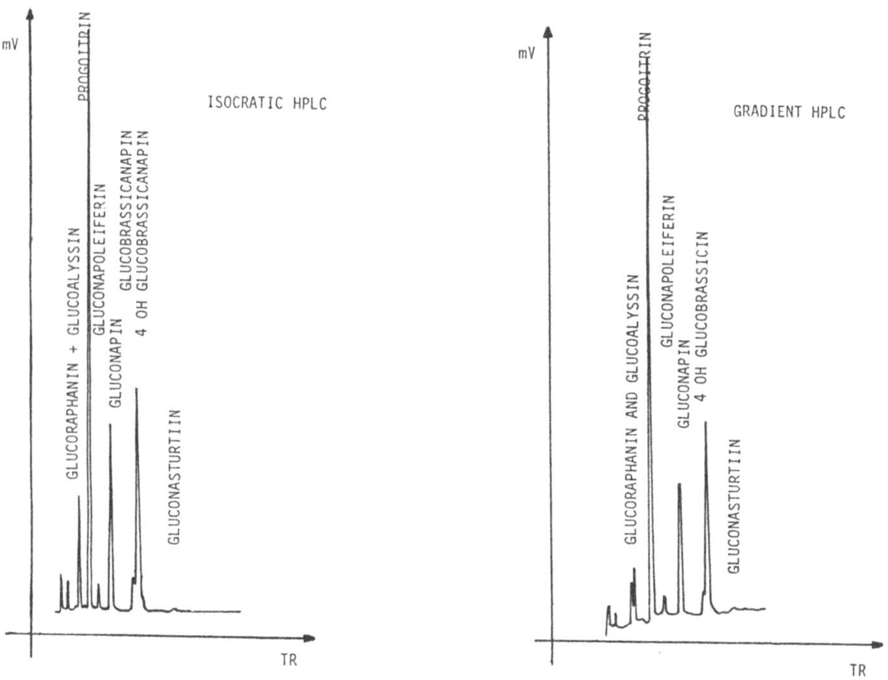

Fig. 8 : HPLC of intact glucosinolates
(acetonitrile buffer as eluant)

The results obtained with this new acetonitrile buffer have been realised in Sørensen laboratory.

On the other hand, by Sørensen's method, it is possible to detect and measure derivatives of glucosinolates (cinnamoyl derivatives...) found in some seeds. It should be very interesting to have the possibility of extracting these products and to follow their evolution by using Spinks method (Spinks et al.,1983). Are these products toxic and are they interfeering with the measurement of total glucosinolates ? Research should be done to answer these questions.

In a general way, we think, from a qualitative point of view that Sørensen method is interesting and complementary to the desulphoglucosinolate method. The use of both technics should avoid some interpretation errors.

Concerning retention times, we have not achieved such good reproducibility using methanol buffer.

CONCLUSIONS

As general conclusions, we can say that it is not necessary to defat the seeds before glucosinolate extraction with aqueous methanol.

D.E.A.E. sephadex A 25 columns of 15, 25 and 35 mg used to purify the desulphoglucosinolates give similar results, which are not the same for Ecteolla cellulose columns.

The pyridin acetate buffer used for the preparation of D.E.A.E. columns can be usefully replaced by an acetic acid- sodium acetate buffer at pH 5.

By GC-MS, we have confirmed the presence of gluconapin, gluoobrassicin, glucobrassicanapin, progoitrin, gluconapoleiferin and 4 OH glucobrassicin in Garant, Ledos, Librador and Lirasol varieties.

From a qualitative point of view, the use of the two methods (desulphoglucosinolates and intact glucosinolates) at the same time is interesting in some cases to highlight glucosinolates that could have the same retention time with only one method.

Glucotropaeolin and glucobarbarin are good internal standards and it would be very convenient to be able to commercialize them.

As well as proposing some methods to determine the molar absorption coefficients, efforts must be made to obtain ultra pure glucosinolates.

The temperature of desulphoglucosinolate HPLC column must be set with precision; an increase in temperature means a loss of 4 OH glucobrassicin. Complementary experiences should prove if this loss is due to a degradation of the compound in the column.

Spherisorb columns have a long life and the retention times are very reproducible.

The desulphoglucosinolates measurements by GLC and HPLC are done in routine in our laboratory.

Research must keep going on with the purpose of testing acetonitrile buffer used to separate intact glucosinolates. This new buffer separates the glucosinolates in a better way and decreases the risk of high column pressure.

In a future work, the glucosinolates derivatives we have found by Sørensen method will be isolated to follow their evolution during desulphation and to check their influence on total glucosinolate determination.

ACKNOWLEDGEMENTS

Support from CEC is gratefully acknowledged

REFERENCES

Christensen, B., Kjaer, A., Madsen, Ø, Olsen, C., Olsen O. and Sørensen ,H. 1982. Mass-spectrometric characteristics of some pertrimethyl-silylated desulphoglucosinolates. Tetrahedron, 38 , 353-357 .

Eagles, J., Fenwick, G.R., Gmelin, R. and Rakow,D.1981 . The chemical ionization mass spectra of glucosinolates (Mustard Oil Glycosides) and desulphoglucosinolates. A useful aid for structural analysis. Biomedical mass spectrometry, 8 , 265-269.

Heaney, R.K., Spinks, E.A., Hanley,A.B.and Fenwick, G.R. 1986. Analysis of glucosinolates in Rapeseed. Technical bulletin. Agricultural and Food Research Council. Food Research Institute, Norwich, 1-25.

Helboe, P., Olsen, O. and Sørensen, H. 1980. Separation of glucosinolates by high performance liquid chromatography. Journal of Chromatography, 197 , 199-205.

Møller, P., Olsen, O., Plöger, A., Rasmussen, K. and Sørensen, H. 1985. Quantitative analysis if individual glucosinolates in double-low oilseed rape by HPLC of intact glucosinolates. In : Advances in the production and utilisation of cruciferous crops (ed. Sørensen,H.) Nijhoff/Junk; Dorbrecht, Boston and Lancaster, 111-126 .

Spinks, E., Sones, K. and Fenwick,G.R. 1984 . The quantitative analysis of glucosinolates in cruciferous vegetables, oilseeds and forage crops using high performance liquid chromatography. Fette, Seifen,Anstrichmittel, 86 , 228-231.

QUANTITATIVE ANALYSIS OF GLUCOSINOLATES IN OILSEED RAPE BASED ON HPLC OF DESULFOGLUCOSINOLATES AND HPLC OF INTACT GLUCOSINOLATES

Birthe Bjerg and Hilmer Sørensen

Chemistry Department,

Royal Veterinary and Agricultural University,

40, Thorvaldsensvej, DK-1871 Frederiksberg C, Denmark.

ABSTRACT

High performance liquid chromatography (HPLC) of desulfoglucosinolates and of intact glucosinolates are the only two methods of analysis known at present, which can fulfil the requirements for reliable determinations of individual glucosinolates in oilseed rape and other Brassica species or cultivars. Use of HPLC requires purification of the glucosinolate-containing extract to avoid serious problems from interfering compounds. Analysis of glucosinolates in double low oilseed rape or other kinds of materials with a low content of glucosinolates requires in addition a concentration of the sample. The experimental steps in performance of these two types of analysis are nearly identical.

The requirements for the extraction of glucosinolates for preparation of 'crude glucosinolate extracts' do not deviate for the two methods. It is recommended to use a glucosinolate of known purity as internal standard and added to the extraction solution before start of the homogenisation to control the myrosinase inactivation efficiency. It is also recommended to perform the analysis with and without internal standard or with different glucosinolate standards in the two determinations, e.g. glucotropaeolin and glucobarbarin. The desulfotechnique requires concentration of the extract to remove the methanol; an evaporation step is also acceptable, but not required, for the technique based on intact glucosinolates.

Purification and concentration are in both techniques based on use of mini-columns. The differences consist in the column material used and the elution step where desulfatation with sulfatase or elution with 0.1 M hydrogencarbonate (or 0.1 M hydrogenphosphate, pH 8.0) are used in the techniques based on desulfoglucosinolates or intact glucosinolates, respectively.

The HPLC steps are in both cases based on reversed phase columns of the same types. The mobile phases are for both methods acetonitrile - water mixtures which for the intact glucosinolates in addition require a counterion content. HPLC of desulfoglucosinolates is a gradient technique which requires two pumps and a system controller. HPLC of intact glucosinolates can be performed both as a gradient technique and an isocratic technique which is based on a simple instrumentation with only one pump. A column heater is recommended for both techniques. The same types of autosampler and laboratory integrator or microcomputer can be used with advantage in both techniques.

Both of the HPLC techniques give a nearly equal and high resolution of the peaks with use of the methods now recommended. These consist in minor modifications of the original proposals for HPLC of glucosinolates. However, for the intact glucosinolates, the reduced viscosity of the mobile phase obtained by change of the mobile phases composition and especially by using 50°C or 70°C as column temperature has improved the technique appreciably.

Limitations and possibilities as well as advantages and disadvantages of these two techniques are presented and discussed briefly. These two HPLC methods supplement each other very well, the results obtained from both methods can be used either alone or in combination. It is recommendable to have both methods as accepted standard methods. Thereby, different laboratories according to their available instrumentation and HPLC experience can choose, without problems, the method or methods which fulfil their requirements.

INTRODUCTION

Double low oilseed rape varieties are promising crops which primarily are used as a source of vegetable oil and protein (Larsen and Sørensen, 1985). The rapeseed proteins have a well balanced amino acid composition and consequently a high nutritive value (Eggum et al., 1985a and 1985b; Andersen and Sørensen, 1985). However, a dominating problem limiting an optimal utilization of oilseed rape is the possibility for a too high content of glucosinolates or degradation products thereof in the oil and meal products (Eggum et al., 1985c).

Solution of the problems caused by glucosinolates requires reliability of the applied analytical methods (Sørensen, 1985). HPLC of desulfoglucosinolates (Minchinton et al., 1982; Spinks et al., 1984) and HPLC of intact glucosinolates (Helboe et al., 1980; Møller et al., 1985) are the only available methods of analysis for individual glucosinolates which fulfil the requirements.

The purpose of the present investigations has been a comparison and harmonization of these two HPLC methods. The described extraction procedure has been developed as a simple quantitative method of the crude extract preparation. This extract can be used in both ion-exchange purifications required prior to the HPLC analysis of the intact and of desulfoglucosinolates, respectively. The paper supplements previously presented papers on this subject (Spinks et al., 1984; Møller et al., 1985). Thereby, some improvements of especially the resolution and shape of the peaks in the HPLC-chromatograms of intact glucosinolates have been obtained. In addition, HPLC of intact glucosinolates based on gradient elution, requirements to response factors and use of internal standards, effects of changes in chromatographic conditions and possibilities of autosampler applications are described.

MATERIALS AND METHODS

General methods and instrumentations used at the majority of the different experimental steps have been presented elsewhere. In the text refers one asterisk (*) to details in the paper by Møller et al. (1985) and two asterisks (**) refer to details in the paper by Spinks et al. (1984).

Homogenization and extraction; "crude glucosinolate extract"

Apparatus and reagents as previously described*.

1. Weigh out ca. 0.5 g of sample (seeds or meal) with an accuracy of $\pm$0.1 mg in a centrifuge tube.
2. Add 3 ml boiling extraction solution (and internal standard, vide infra)* and homogenize keeping the centrifuge tube in a boiling water bath*.
3. Centrifuge, transfer the supernatant to a volumetric (10 ml) flask, repeat twice the homogenization and extraction of the sediment, and make up the extract of the "crude glucosinolate extract" to the 10 ml mark*.
4. If required, e.g. in the desulfoglucosinolate methods of analysis or with very low concentrations of glucosinolates in the samples, the "crude glucosinolate extracts" can be concentrated by evaporation to 1-2 ml (not to dryness) and diluted with water to a known volume giving "crude glucosinolate extract in water".

Isolation of intact glucosinolates

1. Prepare Ecteola-Cellulose in the acetate form*
2. Arrange mini-columns (plastic pipette tips with plugs of glass wool as bottom).

3. Use 1.0 ml of the Ecteola material (an equal volume of column material and water) to each column and wash with water*

4. Transfer 1000 µl "crude glucosinolate extract" to the top of the column, carefully and without destruction of the plane surface. Allow to drain.

5. Wash with 2 x 0.5 ml water carefully as in (4). Allow to drain.

6. Place a small beaker (or autosampler glass) containing 200 µl 1.5 M hydrochloric acid below the columns.

7. Elute with 2.8 ml sodium hydrogencarbonate (0.1 M, pH 9.0) or with 2.8 ml phosphate buffer (0.1 M, pH 8.0). For both buffer solutions, the highest obtainable purity is required to avoid peaks in the HPLC chromatogram from impurities.

8. Mix the 3.0 ml eluate containing the isolated intact glucosinolates. It is ready for analysis or can be stored frozen until required.

Isolation of desulfoglucosinolates

1. Prepare DEAE Sephadex A25 suspension and sulphatase enzyme solution**.

   Stir DEAE Sephadex A25 powder (Pharmacia) in excess acetate buffer (see 6). Filter and resuspend in fresh buffer. Filter and wash the Sephadex with water, then resuspend in the acetate buffer to give a total volume twice that of the settled resin. Dissolve 70 mg of sulphatase (Type H-1, Sigma) in 3 ml water. Add 3 ml of ethanol, centrifuge and discard the precipitate. To the supernatant, add 1.5 vol ethanol, centrifuge and dissolve the precipitate in 2 ml water. Pass the solution in turn through small columns prepared from 20-30 mg dry weight of DEAE Sephadex A25 (acetate form) and SP Sephadex C25 (sodium form, Pharmacia). The resulting solution should contain approximately 0.3 U enzyme/ml. Store at -18°C until required.

2. Arrange mini-columns (plastic pipette tips with plugs of glass wool as bottom)** or *

3. Use 1.0 ml of the DEAE Sephadex A25 suspension (an equal volume column material and acetate buffer) to each column and wash with water**

4. Transfer 1000 µl "crude glucosinolate extract" to the top of the column, carefully and without destruction of the plane surface. Allow to drain.

5. Wash with 2 ml water as in (4). Allow to drain.

6. Wash with 2 x 0.5 ml 0.02 M acetate buffer, pH 5.0 (1.2 ml acetic acid + 990 ml H_2O + NaOH to pH 5.0 + water to a total volume of 1000 ml). Allow to drain.

7. Place a small beaker (or autosampler glass) below the column.

8. Add 75 µl of sulphatase solution to the top of the column.

9. Leave 16-18 h.

10. Elute with 3 x 0.5 ml water, draining the column between each addition.

11. Mix the 1.5 ml eluate containing the isolated desulfoglucosinolates. It is ready for analysis or can be stored frozen until required.

High Performance Liquid Chromatography (HPLC)

Apparatus: HPLC equipment and other apparatus as previously described*; gradient elution requires two pumps and system controller units, for the isocratic system one pump is sufficient. Column heater, autosampler and laboratory integrator or microcomputer.
Reagents: Acetonitrile, counterions, water and other solvents and reagents used for especially the gradient HPLC technique must be of very high purity required for HPLC and degassed before use; other reagents as previously*.
HPLC columns: (A) Nova Pak ^{TM}C-18, 5 micron (150 x 3.9 mm i.d.) (Waters-Millipore A/S), (Ax) Spherisorb, 3 micron (150 x 4.6 mm i.d.) (Phase Separations Inc.) or Nucleosil$^{(R)}$ C-18, 5 µm (Macherey-Nagel) packed by the dilute slurry technique in column (B)(150 x 4.6 mm i.d.) or column (C)(250 x 4.6 mm i.d.).

Gradient HPLC of desulfoglucosinolates; the mobile phases are:

(A) Water; redistilled and/or double ion-exchanged and specially purified for all organic impurities

(B) Acetonitrile, 20% (v/v) in (A)

Gradient table:

Flow(ml/min)	Time (min)	A%	B%	Curve
1.5	start	99	1	linear
1.5	1	99	1	"
1.5	21	1	99	"
1.5	24	1	99	"

1.5	29	99	1	"
1.5	39	99	1	"

Detection wavelength 230 nm; column oven 30°C (alternative 50°C or 70°C; vide infra).

<u>Gradient HPLC of intact glucosinolates</u>; several possibilities can be selected depending on the separation requirements.

Two different sets of conditions are shown below and others are presented in the text (vide infra). The mobile phases are:

(A) 0.01 M phosphate buffer (pH 7.0) containing tetra-heptylammonium bromide (5 mM) and modified with 20% acetonitrile

(B) 0.01 M phosphate buffer (pH 7.0) containing tetra-heptylammonium bromide (5 mM) and modified with 40% acetonitrile

Gradient table

Flow (ml/min)	Time (min)	A%	B%	Curve
1	start	99	1	*
1	5	99	1	11
1	40	35	65	6
1	60	35	65	11
1	65	99	1	2

For fast gradient HPLC use the mobile phase A with 25% acetonitrile, B with 60% acetonitrile and the following gradient table

Flow (ml/min)	Time (Min)	A%	B%	Curve
1	start	85	15	*
1	40	60	40	6
1	45	60	40	11
1	47	85	15	2

Detection wavelength 235 nm; column oven as for isocratic HPLC.

<u>Isocratic HPLC of intact glucosinolates</u>: the mobile phase is: 0,01 M phosphate buffer (pH 7.0) containing tetraheptylammonium bromide (5 mM) and modified with 30% acetonitrile. For fast analysis the modifier concentration is increased to 32%. Column temperature at 50°C (alternative 30°C or 70°C, see chromatograms; vide infra). Detection wavelength 235 nm,

HPLC analysis of individual intact glucosinolates

1. Start of the HPLC instrument, equilibrate with mobile phase as required for the isocratic system or the starting conditions for the gradient system described above.

2. Inject a sample of reference glucosinolates (20 µl or by use of autosampler), e.g. rapeseed glucosinolates isolated as described elsewhere*, to control the correct HPLC conditions as expected from reference chromatograms* (vide infra).

3. Perform the analysis of the extracts with isolated intact glucosinolates.

4. Use areas of the peaks corresponding to the different glucosinolates* and the area of the internal standards (with known amount (and purity) added per g of seed), or use external standard (standard curve), for the calculations*. Corrections by use of response factors are required (vide infra).

HPLC analysis of individual desulfoglucosinolates

1. Start the HPLC instrument, equilibrate with the mobile phases and starting conditions for the gradient system described above.

2. Inject a sample of reference desulfoglucosinolates (20 µl or by use of autosampler), e.g. a sample of rapeseed glucosinolates isolated as described elsewhere* and transformed into desulfoglucosinolates on mini columns (vide supra), to control the correct HPLC conditions as expected from reference chromatograms (vide infra).

3. Perform the analysis of the extracts with isolated desulfoglucosinolates.

4. Use areas of the peaks corresponding to the different desulfoglucosinolates (vide infra) and the area of the internal standards (with known amount (and purity) added per g of seed), or use an external desulfoglucosinolate standard (standard curve), for the calculations as described for intact glucosinolates*. Corrections by use of response factors are required (vide infra).

Calculation of results, use of reference compounds and response factors

Quantitative determinations using chromatographic peak areas require that we include corrections for the different absorptivities of the various compounds at the wavelength used. Such provisional response factors have been determined for several intact glucosinolates at the detection wavelength (235 nm) used in the described HPLC technique*. Other response factors are required for desulfoglucosinolates** at the detection wavelength used (230 nm). For both HPLC methods other response factors are required if other wavelengths and temperatures are used (vide infra).

It is recommended to express the results in µmole/g of air dried seed (7% water) or in µmole/g of the dry material used (e.g. rapeseed meal). Calculations can be based on external standards (standard curves)* or internal standards. In the latter case, internal standards with retention times different from all of the glucosinolates present in the samples are necessary. Glucotropaeolin and/or glucobarbarin can be recommended as internal standards for both of the described HPLC analysis of glucosinolates*. If compounds in the samples used for analysis have retention times as the internal references, the problem can quite often be solved by changing the column temperature to e.g. 50° or 70°C. The problems are revealed by performing the analysis +/- internal standards or by using glucotropaeolin in one trial and glucobarbarin in the other.

The internal standard must be a glucosinolate, introduced before homogenization, because an insufficient myrosinase inactivation is one of the most critical steps in all methods of glucosinolate analysis. In analysis of samples containing few to about 20 µmole of glucosinolates/g (e.g. double low oilseed rape), it is recommended to use 2 µmole internal standard/g (250 µl of a 4 mM solution/0.5 g).

For the calculations based on internal standard (I.S.):

µmole glucosinolate /g sample =

$$\frac{\text{area of glucosinolate}}{\text{area of I.S.}} \times \text{response factor} \times \text{µmole I.S./g sample}$$

Results are given as mean values for at least two, of each

other independent, determinations with indication of the deviation.

RESULTS AND DISCUSSION

The method used for preparation of the "crude glucosinolate extract" from oilseed rape and rapeseed meal has been shown to result in quantitative extraction of the glucosinolates. This is revealed from the recovery obtained by use of several reference glucosinolates (Bjerg and Sørensen, 1986). The glucosinolates in the methanol-water extract stored in a refrigerator are stable for several days. However, in extracts containing some metals in appreciable amounts (especially likely for some food and feedingstuffs) a possibility exists on non-enzymatic glucosinolate degradation (Olsen and Sørensen, 1981).

Intact glucosinolates (or most of them) isolated by the described extraction technique and stored in a refrigerator are stable for several weeks if microbial growth is excluded. Exceptions are the glucosinolates which are oxidatively and hydrolytically sensitive/unstable (Bjerg and Sørensen, 1986). This seems also to be the case for desulfoglucosinolates prepared by the described technique. Pyridine as well as neutral to alkaline conditions in the solutions containing the unstable glucosinolates and desulfoglucosinolates are not acceptable for longer periods.

For both of the ion-exchange methods it is important to keep the now recommended column size and elution volume in accordance with previously described results for the intact glucosinolates (Castor-Normandin et al., 1986). Thereby, it is possible to obtain quantitative isolation, sufficient purification and concentration of all known intact glucosinolates present in "crude glucosinolate extracts" from oilseed rape and rapeseed meal. This is also the case for most of the known glucosinolates when the desulfotechnique is used (Sørensen, 1985; Bjerg and Sørensen, 1986).

The possibilities of optimal HPLC separation and quantitative determination of individual glucosinolates/desulfoglucosinolates require consideration of both the structure and properties of relevant glucosinolates (Fig. 1 and Table 1) and of the applied chromatographic materials.

Details about the structure and properties of the Ecteola-Cellulose (Bjerg et al., 1984) and of glucosinolates and their degradation products have been presented and discussed elsewhere (Olsen

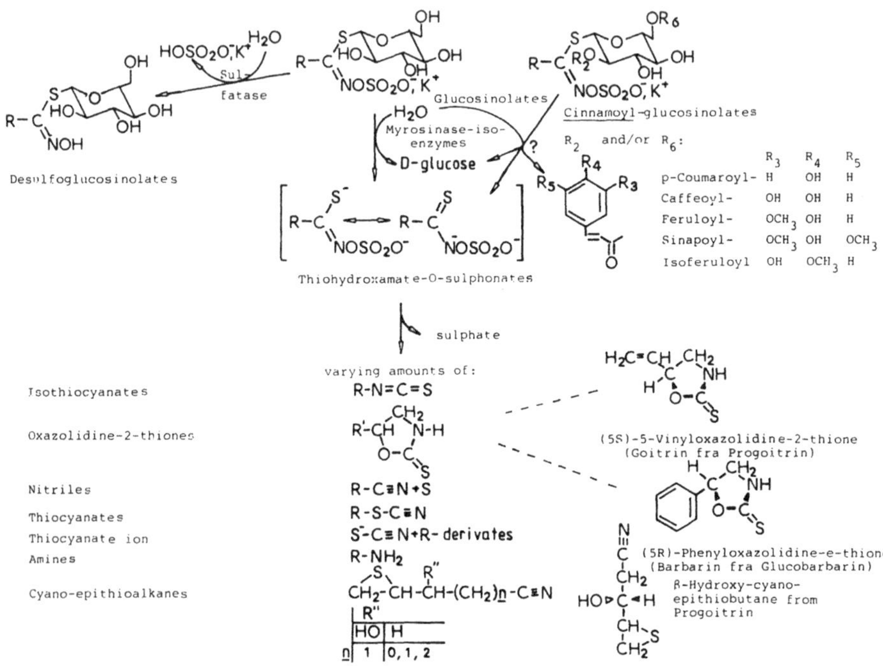

Fig. 1. Names and structures of degradation products of glucosinolates formed by sulfatase- and myrosinase-catalysed hydrolysis: R = side chain of glucosinolates; - structures and names of some selected glucosinolates are shown in Table 1. R_2 and/or R_6 are cinnamoyl-derivatives.

and Sørensen, 1981; Bjerg and Sørensen, 1986; and refs. cited therein). For an efficient utilization of HPLC in analysis of glucosinolates and desulfoglucosinolates, we need to realize the properties of the compounds as well as advantages and disadvantages of the properties of the chromatographic materials and mobile phases required as a result thereof.

HPLC of both desulfoglucosinolates and intact glucosinolates are based on reversed phase HPLC and for the latter ion-pairing is required. As a result thereof, hydrophilic solvents are required and thereby increased viscosity leading to reduced separation effi-

TABLE 1. Selected glucosinolates representing structures which need to be considered in discussions of quality of oilseed rape and different cruciferous materials and thereby requirements to methods of glucosinolate analysis.

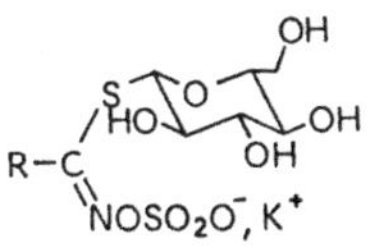

and

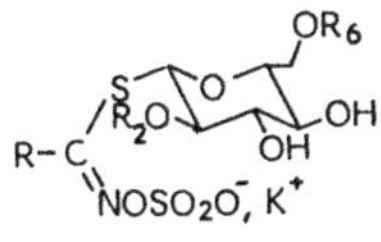

R_2 & R_6 = cinnamoyl-derivatives (Fig. 1)

Glucosinolates

No.	Structure of R group	Semisystematic names	Trivial names
1	$CH_2=CH-CH_2-$	Allylglucosinolate	Sinigrin
2	$CH_2=CH-CH_2-CH_2-$	But-3-enylglucosinolate	Gluconapin
3	$CH_2=CH-CH_2-CH_2-CH_2-$	Pent-4-enylglucosinolate	Glucobrassicanapin
4	$CH_2=CH-CH(OH)-CH_2-$	(2R)-2-Hydroxybut-3-enylglucosinolate	Progoitrin
5	-- " --	(2S)-2-Hydroxybut-3-enylglucosinolate	Epiprogoitrin
6	$CH_2=CH-CH_2-CH(OH)-CH_2-$	(2R)-2-Hydroxypent-4-enylglucosinolate	Napoleiferin
7	$CH_3-S-CH_2-CH_2-CH_2-$	3-Methylthiopropylglucosinolate	Glucoibervirin
8	$CH_3-S-CH_2-CH_2-CH_2-CH_2-$	4-Methylthiobutylglucosinolate	Glucoerucin
9	$CH_3-S-CH_2-CH_2-CH_2-CH_2-CH_2-$	5-Methylthiopentylglucosinolate	Glucoberteroin
10	$CH_3-SO-CH_2-CH_2-CH_2-$	3-Methylsulphinylpropylglucosinolate	Glucoiberin
11	$CH_3-SO-CH_2-CH_2-CH_2-CH_2-$	4-Methylsulphinylbutylglucosinolate	Glucoraphanin
12	$CH_3-SO-CH_2-CH_2-CH_2-CH_2-CH_2-$	5-Methylsulphinylpentylglucosinolate	Glucoalyssin
13*	$CH_3-SO-CH=CH-CH_2-CH_2-$	4-Methylsulphinylbut-3-enylglucosinolate	Glucoraphenin
14	$CH_3-SO_2-CH_2-CH_2-CH_2-$	3-Methylsulphonylpropylglucosinolate	Glucocheirolin
15	$CH_3-SO_2-CH_2-CH_2-CH_2-CH_2-$	4-Methylsulphonylbutylglucosinolate	Glucoerysolin
16	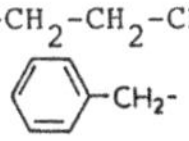	Benzylglucosinolate	Glucotropaeolin
17	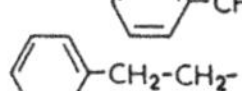	Phenethylglucosinolate	Gluconasturtiin
18*	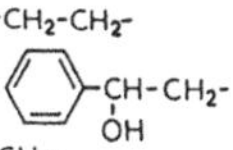	2-Hydroxy-2-phenylethylglucosinolate	Glucobarbarin
19	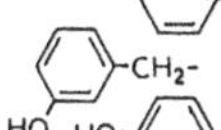	m-Hydroxybenzylglucosinolate	Glucolepigramin
20	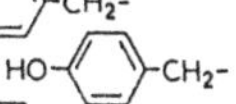	p-Hydroxybenzylglucosinolate	Sinalbin
21	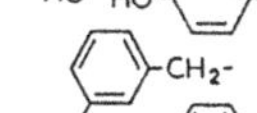	m-Methoxybenzylglucosinolate	Glucolimnanthin
22	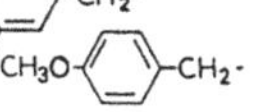	p-Methoxybenzylglucosinolate	Glucoaubrietin
	Indol-3-ylmethylglucosinolates:		
23	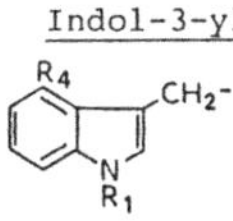 ($R_1=R_4=H$	Indol-3-ylmethylglucosinolate	Glucobrassicin
24	$R_1=OCH_3$; $R_4=H$	N-Methoxyindol-3-ylmethylglucosinolate	Neoglucobrassicin
25	$R_1=SO_3^-$; $R_4=H$	N-Sulphoindol-3-ylmethylglucosinolate	Sulphoglucobrassicin
26	$R_1=H$; $R_4=OH$	4-Hydroxyindol-3-ylmethylglucosinolate	4-Hydroxyglucobrassicin
27	$R_1=H$; $R_4=OCH_3$	4-Methoxyindol-3-ylmethylglucosinolate	4-Methoxyglucobrassicin

The glucosinolates No. 1-15 are biosynthetically derived from methionine. Those derived from phenylalanine (No. 16-22) are of interest as intern references.
*Occur also as cinnamoylderivatives (Fig. 1).

ciency may be a result. This problem can, however, be more or less avoided by an optimized technique, as shown in the following.

Stationary and mobile phases in reversed phase HPLC.

HPLC of intact glucosinolates and the desulfoglucosinolates are based on the reversed phase technique (Helboe et al., 1980; Michington et al., 1982) and with ion-pairing included for the intact glucosinolates. Different column materials and counter-ions can be used (Helboe et al., 1980), and Fig. 2 illustrates a selected example with C-18 material.

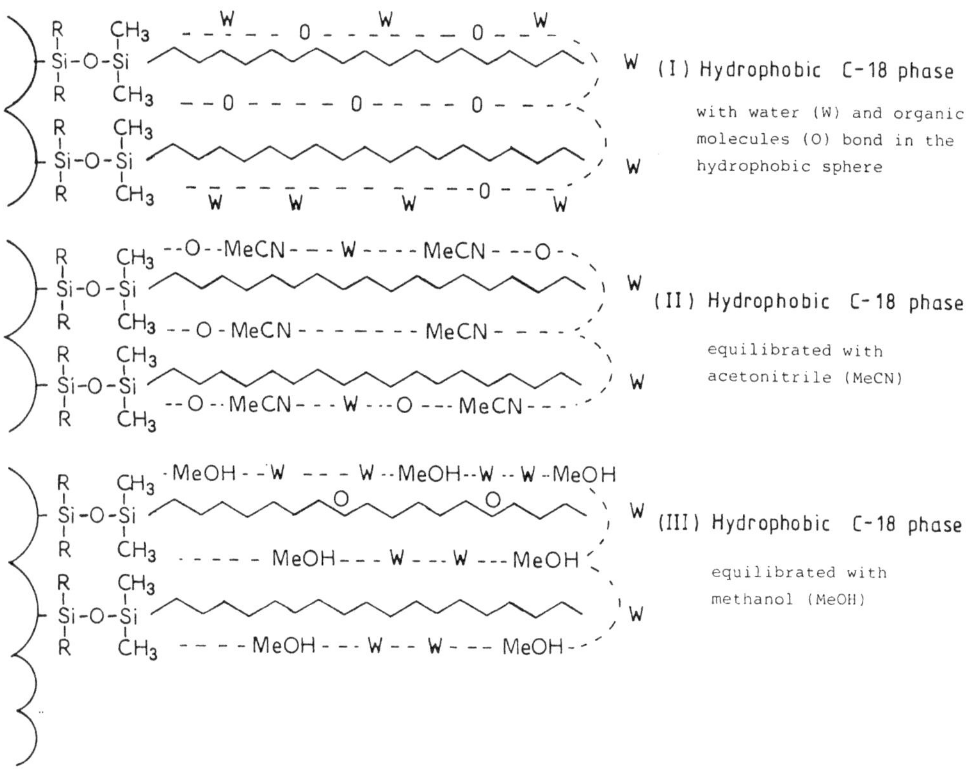

Fig. 2. Illustration of three different conditions (I, II and III) in a reversed phase C-18 stationary phase with quite different characteristic properties as discussed in the text.

The mobile phases used in reversed phase HPLC analysis of glucosinolates and desulfoglucosinolates need to be miscible with water and with a proper polarity. A possible choice may be from the series:

acetone - acetonitrile - isopropanol - methanol - water - acetic acid - formic acid ⟶ polarity

Limitation in the possible choice may be caused by insufficient purity of available solvents and too strongly acidic solvents may cause destruction (hydrolysis) of the stationary phase (Fig.2).

Change from methanol (Møller et al., 1985) to acetonitrile (vide infra) as modifier in the mobile phase reduces the viscosity and thereby improves the peak shape and separation possibilities. Acetonitrile compared to methanol is more efficient in elution of strongly bond compounds as illustrated with organic molecules (O) in Fig. 2.

With increased column temperature, the characteristic properties of the stationary phase are changed, and differently for the different mobile phases, as for other chromatographic systems. Increased temperature has especially a strong effect on strongly bond organic molecules (O; Fig.2). The retention times for glucosinolates and desulfoglucosinolates containing aromatic groups (Table 1) are in accordance with this reduced relatively more than for those with aliphatic R-groups (vide infra). Furthermore, the problems caused by some interfering impurities are reduced by increased temperature.

The mobile phase (A) used in HPLC of desulfoglucosinolates is water, limiting the possibilities of further variations in separations of hydrophilic compounds with retention times close to each other. The mobile phases in HPLC of intact glucosinolates contain counterions (Helboe et al., 1980) which gives additional possibilities for impurities, if the quality of these preparations are not sufficiently high. The problems from such impurities are more pronounced in some of the gradient systems compared to the isocratic system.

Separation of desulfoglucosinolates and intact glucosinolates by HPLC

HPLC of desulfoglucosinolates and HPLC of intact glucosinolates are methods of analysis which efficiently supplement and support each other. It is easy to realize from the above presented discussion that these two methods separate the glucosinolates/desulfoglucosinolates (Fig.1; Table 1) according to the same principle,to some extent differently but with the same trend in retention times for the compounds (vide infra). With the possibility of several closely related glucosinolates, which need to be considered (Table 1), it is also easy to realize that one set of experimental conditions can not fulfil the requirement. From consideration of the HPLC results obtainable and presented by examples in the following chromatograms, it is revealed that both methods have advantages and limitations. If we do not restrict the possibilities of choice of different HPLC conditions, reliable quantitative determination of all glucosinolates can easily be obtained by HPLC.

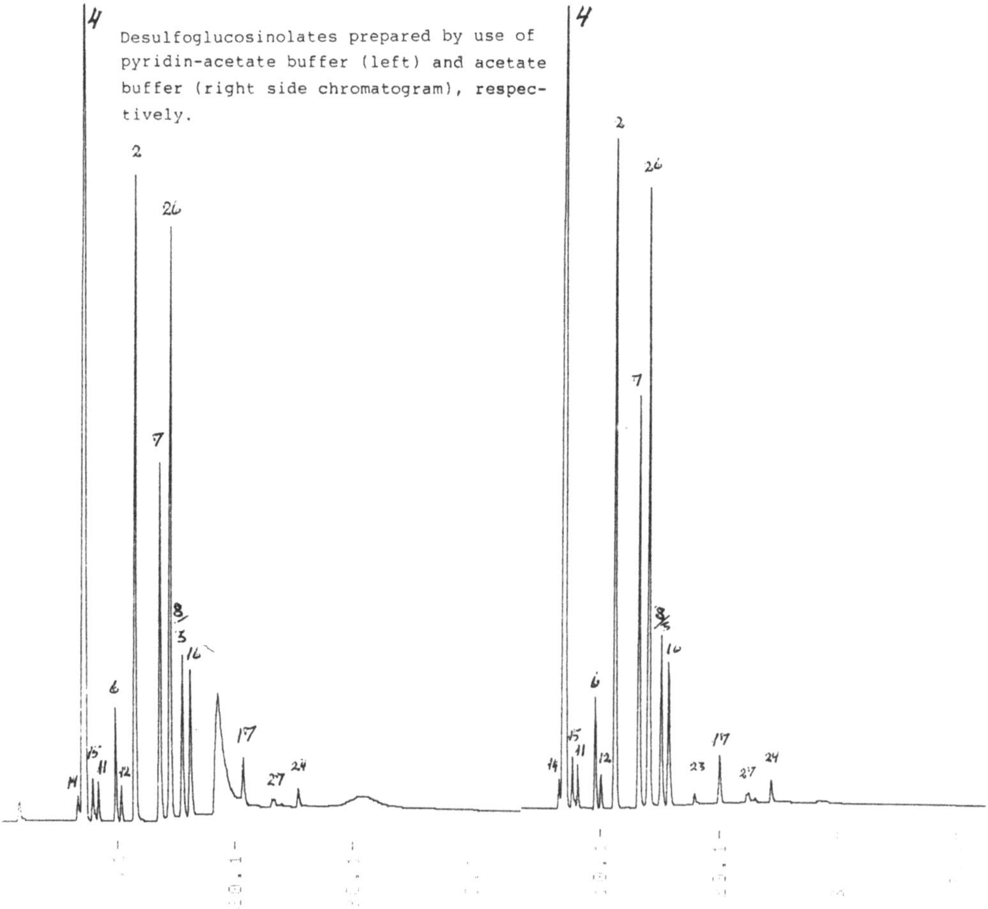

Fig. 3. Separations of desulfoglucosinolates using gradient HPLC, column C,30°C, and detection at 235 nm. The numbers refer to glucosinolates listed in Table 1.

Gradient HPLC of desulfoglucosinolates; separation possibilities, effects of impurities, temperatures and detection wavelength

Fig.3 shows HPLC chromatograms of desulfoglucosinolates. Impurities from pyridine (left side chromatogram) are avoided by the modified procedure (right side chromatogram). It is revealed that some compounds (8 and 3) have not been separated, at the applied conditions.

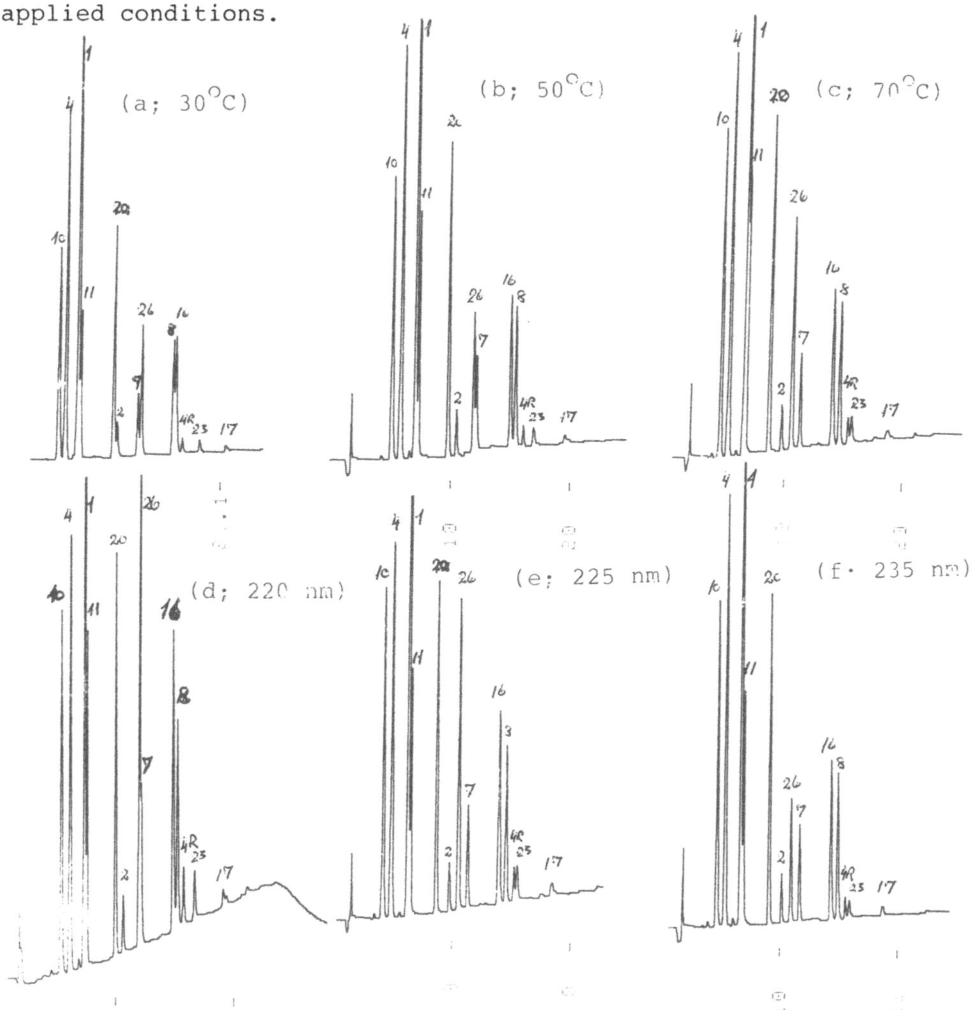

Fig.4. Separation of desulfoglucosinolates using gradient HPLC, column C, different temperatures and different detection wavelength: (a) 30°C and detection at 230 nm; (b) 50°C and detection at 230 nm; (c) 70°C and detection at 230 nm; (d) 50°C and detection at 220 nm; (e) 70°C and detection at 225 nm; (f) 70°C and detection at 235 nm. The numbers refer to glucosinolates listed in Table 1.

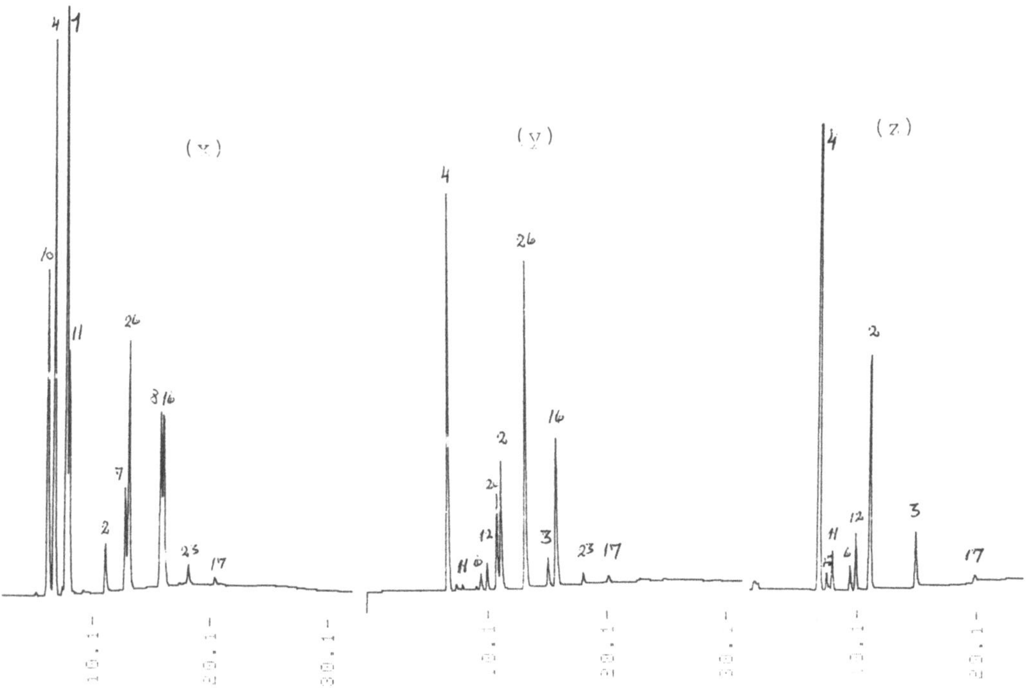

Fig.5. Separation of desulfoglucosinolates using gradient HPLC, column C,30°C, detection at 235 nm and mixtures of desulfoglucosinolates isolated by the described procedures from seeds of three different Brassica varieties (x, y and z).

The temperature affect the retention times (Fig.4; a, b and c) as expected (vide supra). For some compounds (2 and 20, 7 and 26, 8 and 16) improved separations are obtained by increased temperature, whereas the opposite is found for other (1 and 11, 4R and 23).

Fig. 4 shows also the effect of different detection wavelengths used to the same mixture of desulfoglucosinolates. The highest sensitivity is obtained by 220 nm (d), but the effect is different for different desulfoglucosinolates. The problems with impurities and the base line (d) are, however, reduced by 230 nm (c) and 235 nm (f) without serious reduction in the sensitivity.

Fig. 5 shows the possibilities of HPLC separation of desulfoglucosinolate mixtures isolated from seeds of three different Brassica varieties using the procedure now described. It is revealed that very nice chromatograms are obtainable, but is is also revealed that it is not possible to separate all glucosinolates of interest in relation to oilseed rape by use of only one method.

Gradient HPLC of intact glucosinolates; separation possibilities, effects of impurities, temperatures, columns and detection wavelength

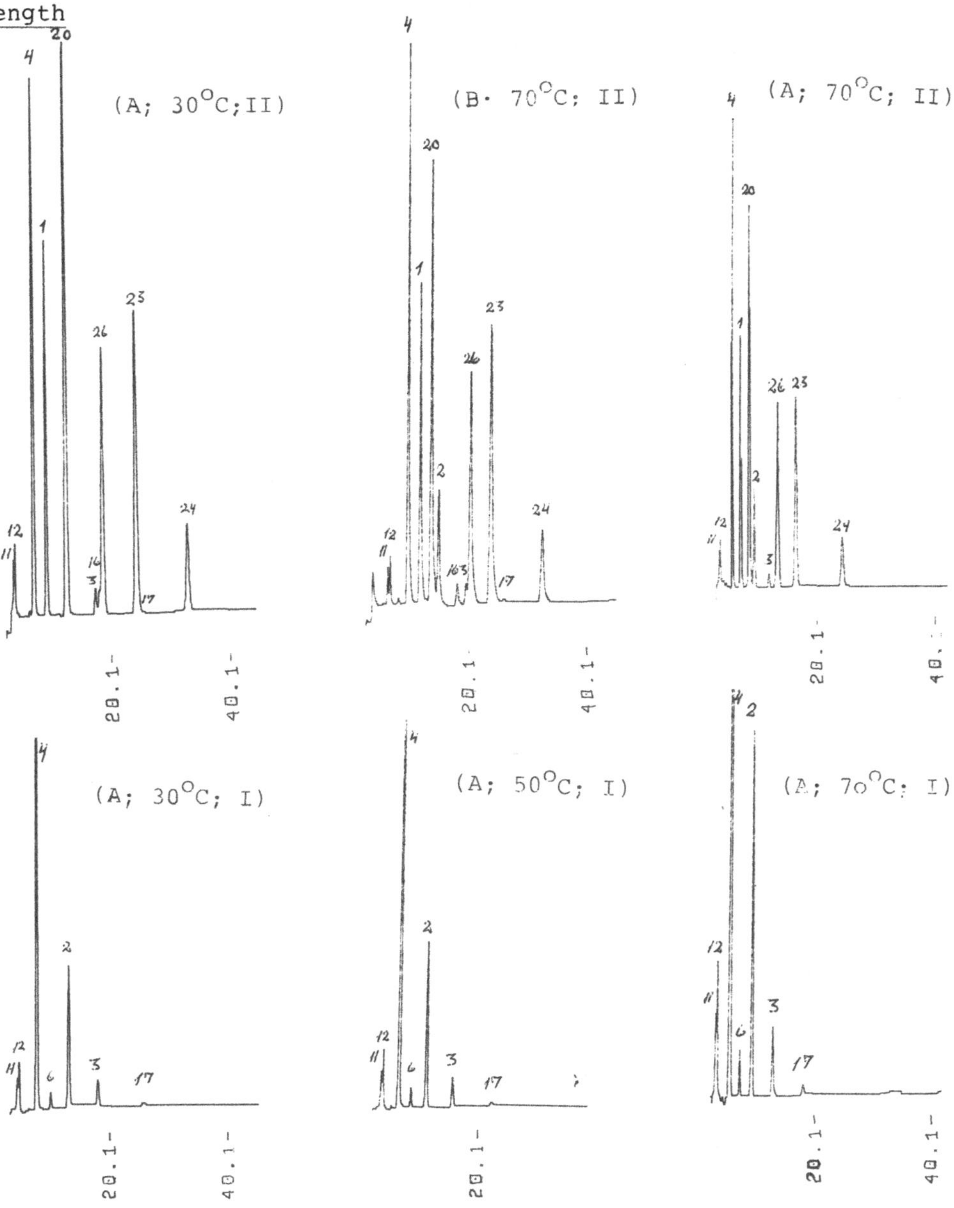

Fig.6. Separation of intact glucosinolates by fast analysis gradient HPLC on column A and B at different temperatures. The mixture I is rapeseed glucosinolates as used previously with methanol as modifier instead of acetonitrile (Møller et al., 1985); mixture II as in fig. 7; numbers refer to glucosinolates shown in Table 1.

The HPLC separation of intact glucosinolates using the modifications now described including the gradient technique (Figs. 6, 7) gives peak shape and trend in retention times for the intact glucosinolates as found for HPLC of desulfoglucosinolates (vide supra). Acetonitrile as modifier instead of methanol reduces the viscosity of the mobile phase and change the surroundings of the hydrophobic phase (vide supra; Fig. 2). This results in improved HPLC separations (Fig. 6) compared with previously obtained results (Møller et al.1985)

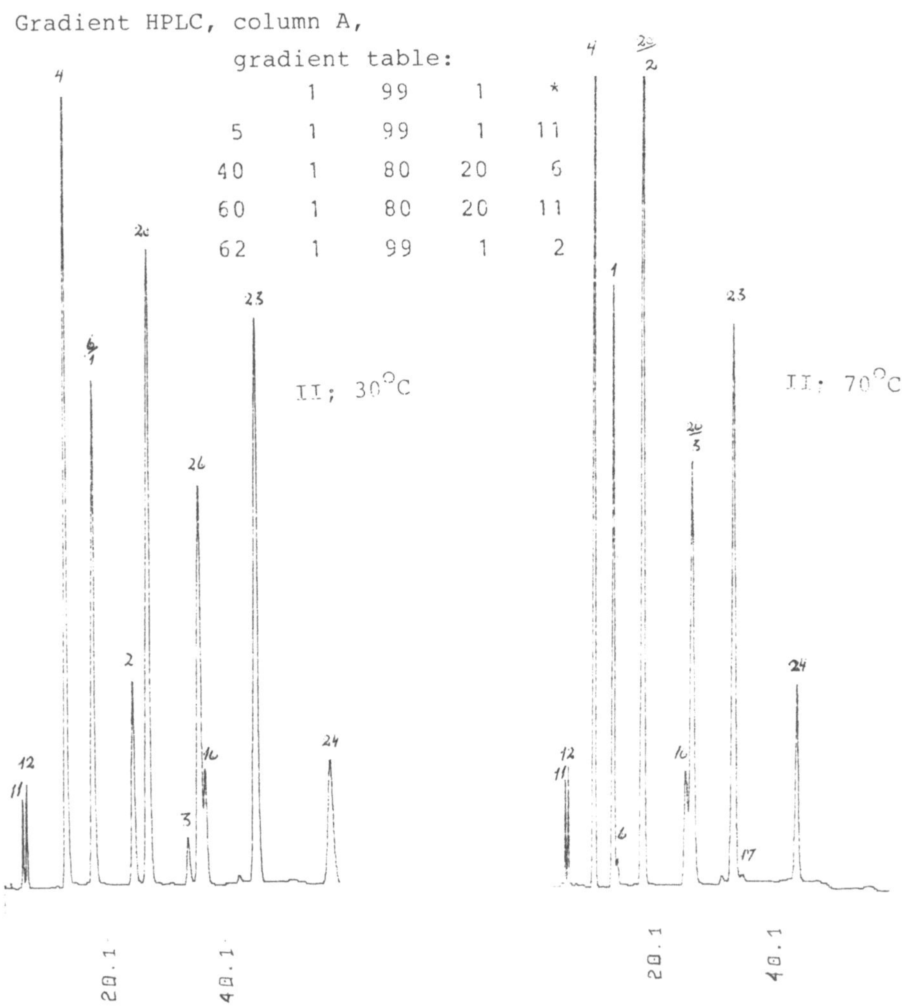

Fig.7. Separation of a mixture of intact glucosinolates (II) by gradient HPLC at different temperatures on column A. The numbers refer to glucosinolates listed in Table 1.

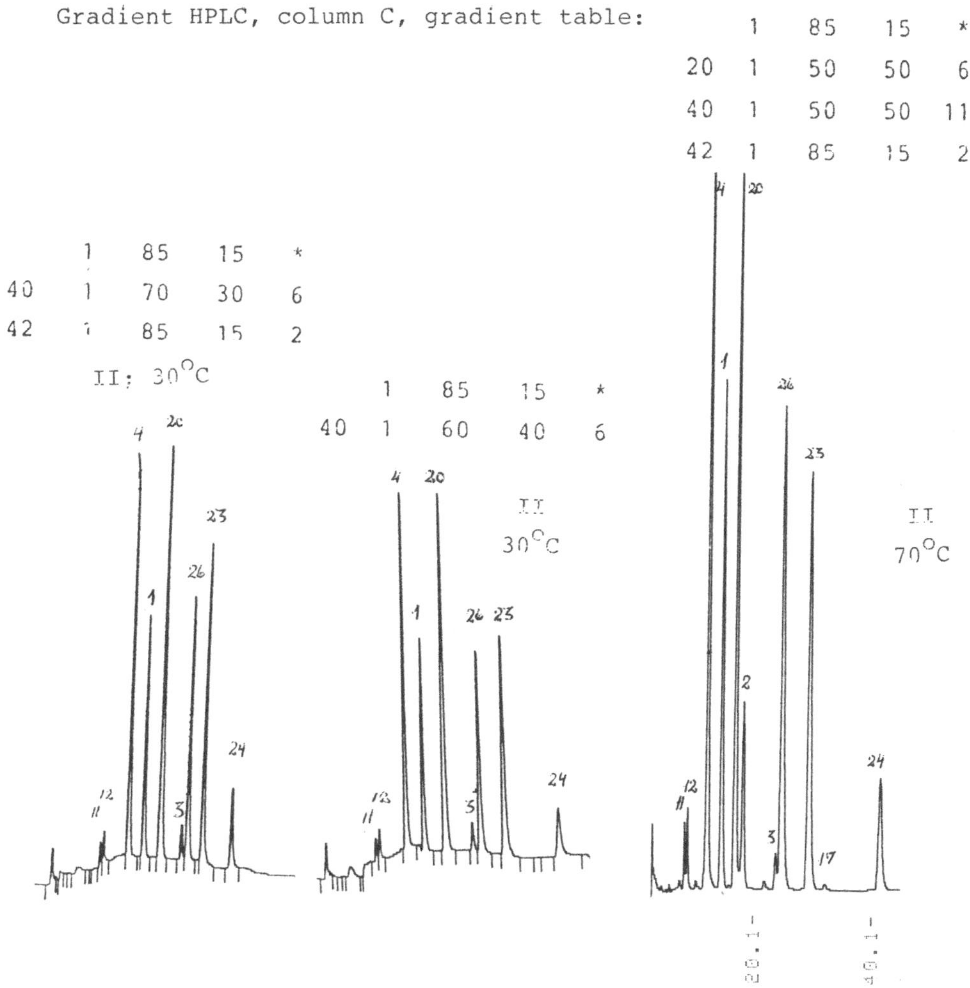

Fig. 8. Separation of intact glucosinolates (see Figs. 6,7) using gradient HPLC on column C.

The increase in column temperature from 30°C to 50°C or 70°C has an even better effect on the viscosity, peak shape and resolution of peaks in HPLC of intact glucosinolates than found for HPLC of desulfoglucosinolates (vide supra). In both cases HPLC at 50°C or 70°C increases the life time for the columns as e.g. problems from organic impurities are reduced as discussed in connection with Fig.2 and the general considerations of reversed phase HPLC.

The results presented (Figs.6-9) show the possibilities of an efficient separation of intact glucosinolates using gradient HPLC at different conditions. Different columns can be used, which again

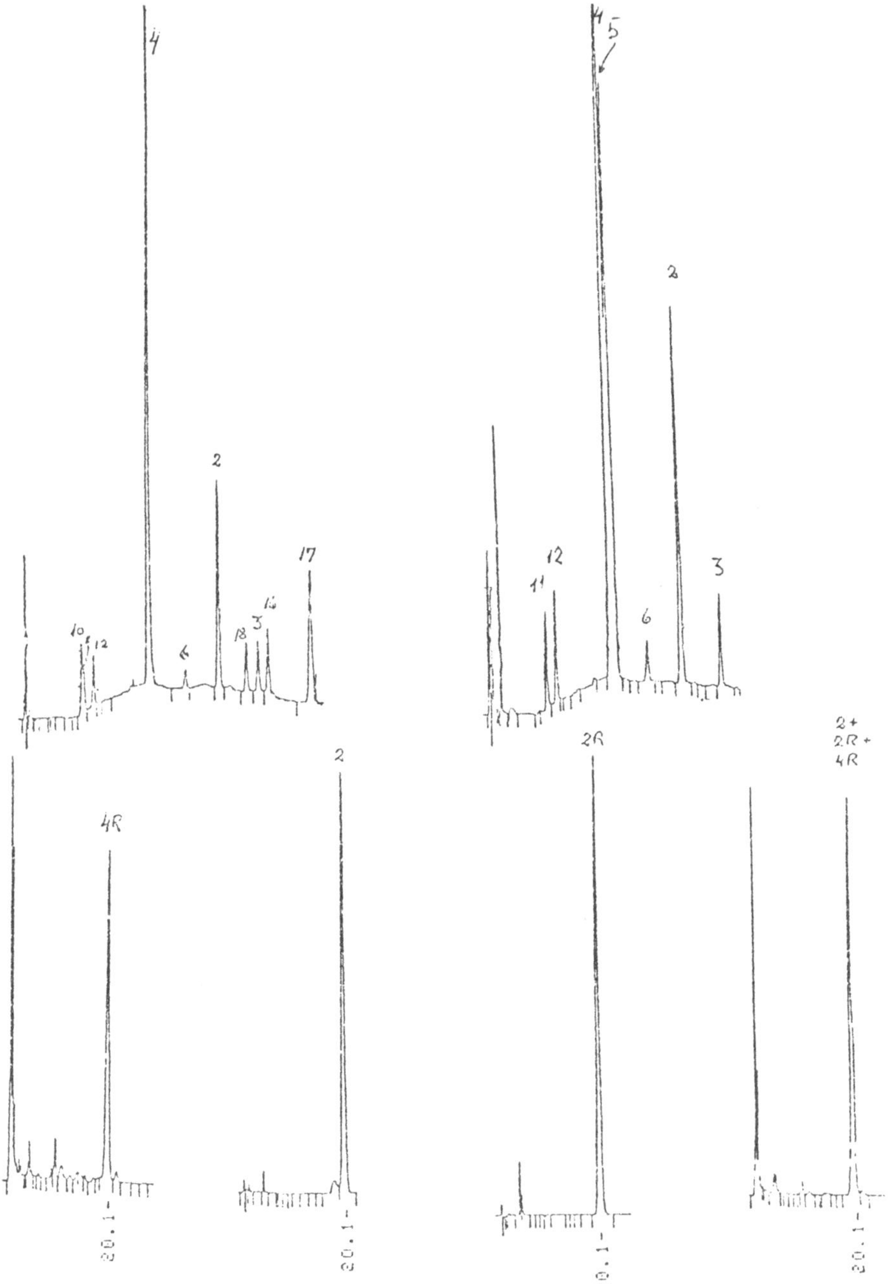

Fig. 9. Separation of intact glucosinolates at 30°C using gradient HPLC on column A. The numbers refer to glucosinolates shown in Table 1. 2R is 2-rhamnopyranosyloxybenzylglucosinolate, 4R is the corresponding p-isomer. The chromatograms at the bottom are of glucosinolates which only have 0.2 min. differences in retention time resulting in one peak of the 1:1:1 mixture.

gives additional possibilities of separation of closely related compounds. There is a high resolution of the peaks, but it is not possible to have base-line separation of all of the more than 100 known different glucosinolates if an acteptable time of analysis is required.

Different types of HPLC conditions are required for separation and reliable determination of all types of glucosinolates. With knowledge to the glucosinolates we want to separate, it will be possible to choose the optimal HPLC conditions from the presented results. Fig. 9 shows that compounds under conditions where the separation of the peaks is small (less than 0.2 min for 4R, 2 and 2R) will result in a single peak if we prepare a 1:1:1 mixture. However, it is also revealed that it is simple to change the HPLC conditions with base-line separation as the result, e.g. for separation of: 2 and 20; 1 and 6; 10, 11 and 12; 18, 3, 16, 26, 23, 17, 27 and 24; and also the diastereoisomers 4 and 5. Gradient HPLC of intact glucosinolates gives great advances in determination of structurally closely related glucosinolates owing to the many possible modifications. However, it will not be possible to select a single se of experimental conditions where all of the possible compounds can be separated, but this is a common problem for all of the known methods.

Impurities in the mobile phases e.g. from the applied water, the modifier and counterion give false peaks or dented base-lines as shown in some of the chromatograms (Figs. 8,9). The dent before the peaks 11 and 12 (Fig.8) is caused by impurities in salts (carbonates or phosphates) used in the buffer solutions.

Isocratic HPLC of intact glucosinolates

For the intact glucosinolates it is possible to use the described isocratic technique with advantages. The (A) column and especially the (Ax) column are recommendable. The effects of changes in the mobile phase, effects of increased column temperature and use of different columns give nearly the same advantages and possibilities as demonstrated and discussed above for gradient HPLC of intact glucosinolates. However, a reduced resolution of peaks from the different glucosinolates by use of isocratic HPLC compared to gradient HPLC of both intact glucosinolates and desulfoglucosinolates have been found. The advantages of isocratic HPLC compared to the gradient technique are an experimental more simple method and a cheaper HPLC system. In addition the problems caused by impurities in the mobile phases are less pronounced when the isocratic technique is used. Figs. 10 and 11 show the separations obtainable by isocratic HPLC used to selected samples of intact glucosinolates.

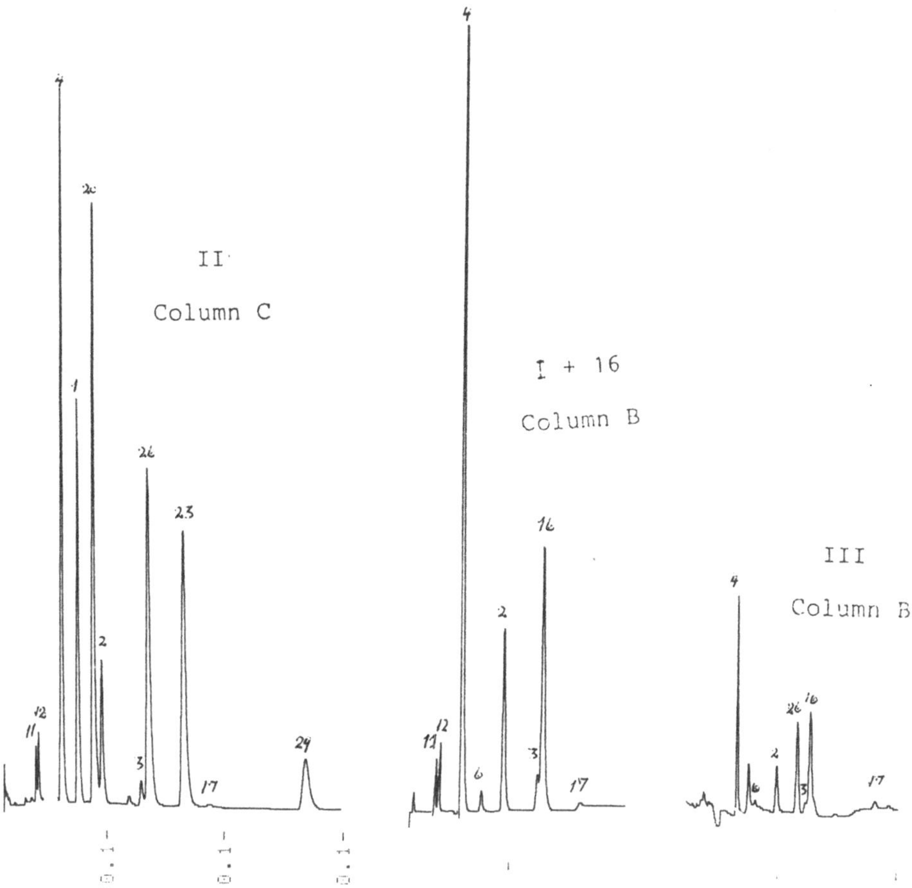

Fig. 10. Separation of intact glucosinolates (see Fig.6 for the mixtures I and II) using isocratic HPLC on column B and C; 70°C; 32% CH_3CN. Mixture III is the glucosinolates isolated as described (vide supra) from a double low rapeseed variety with a low glucosinolate content. Corresponding HPLC results using methanol instead of acetonitrile as modifier have been presented previously (Møller et al., 1985).

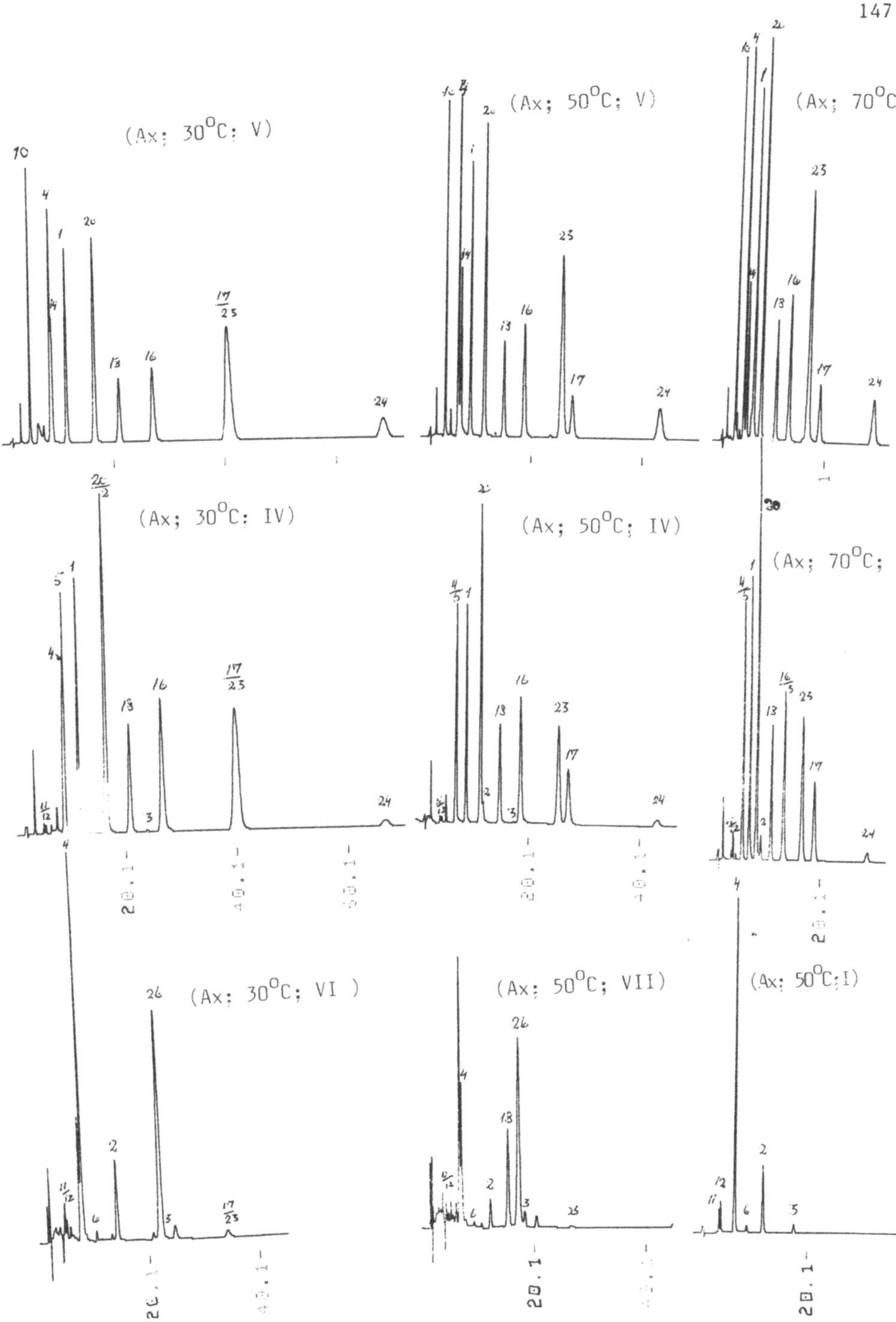

Fig.11. Separation of intact glucosinolates using isocratic HPLC, column (Ax), 30% CH_3CN and different column temperatures as shown on the chromatograms. The numbers at the peaks refer to glucosinolate listed in Table 1. Mixture VI and VII are isolated by the described technique from seeds of two different varieties of double low oilseed rape.

CONCLUSION

The present paper supplements previously reports on this subject (Møller et al., 1985; Sørensen, 1985).

HPLC of desulfoglucosinolates and HPLC of intact glucosinolates are developed as methods of quantitative analysis of individual glucosinolates in oilseed rape and other crucifers.

Both methods require purification and concentration of the crude extract obtained from e.g. seed or meal from double low oilseed rape. This is described in a procedure only deviating at minor points in use of the mini columns. Recommendations with respect to use of internal standards, i.e. intact glucosinolates introduced before start of the extraction, and the requirement to their purity are presented. Unsolved problems concerning those glucosinolates occurring as esters or containing acidic R-groups, when the desulfotechnique is used, have been discussed elsewhere (Bjerg and Sørensen, 1986).

The results presented show the possibilities of an efficient separation of both desulfoglucosinolates and intact glucosinolates using gradient HPLC at different conditions and for the intact glucosinolates isocratic HPLC is also recommendable. There is a high resolution of the peaks, but it is not possible to have baseline separation of all glucosinolates, if only a limited number of HPLC conditions are used. This is a general problem for all three HPLC methods.

Variations in column temperatures have different effects on the retention times of the various desulfoglucosinolates. The temperature effects on desulfoglucosinolates are less pronounced compared to that on intact glucosinolates. However, in both cases the effects are in agreement with changes in the stationary phase as discussed in relation to Fig. 2. Increased column temperature decreases the viscosity of the mobile phases and changes the characteristic properties of the stationary phase resulting in improved separation possibilities and reduced problems from impurities in the mobile phases and/or samples used.

Impurities may result in serious problems with false peaks and/or dented base-lines in chromatograms of both intact glucosinolates (Figs. 8,9) and desulfoglucosinolates (Figs. 3,4). These problems can be solved by using very high purity of buffers, counter-ions and water, avoiding pyridine in the buffer for desulfatation (Fig. 3) and using high column temperature.

The desulfoglucosinolates have UV absorption maxima at 220-230 nm varying for different compounds, and about 5 nm lower compared to the corresponding maxima for the intact glucosinolates. It is recommendable to use 235 or 230 nm as the HPLC detection wavelength to avoid base-line problems caused by impurities in the mobile phases, especially in the case of gradient HPLC. This is not at the UV maxima but acceptable since a limited advance in increased sensitivity at lower wavelengths will be exceeded by the base-line problems caused by impurities.

Response factors have been determined for a comprehensive series of both intact glucosinolates and desulfoglucosinolates at different wavelengths and temperatures. Such response factors are required for reliable determinations using both methods, and they will be presented elsewhere as a result of the EEC supported investigations of these two HPLC methods.

Both HPLC methods can be performed as gradient techniques and for the intact glucosinolates in addition as an isocratic technique. It is impossible to perform separation of all glucosinolates by use of only one method. Results obtained with the glucosinolates included in the chromatograms presented in this paper show that these compounds can be quantitatively determined by use of the described techniques.

It is recommended to have both methods as accepted standard methods since they supplement each other, and results from both methods can be used either alone or in combination. Thereby, different laboratories according to their available instrumentation and HPLC experience can choose, without problems, the method or methods which fulfil their requirements.

ACKNOWLEDGEMENTS

Support from the Danish Agricultural Research Council and from EEC is gratefully acknowledged.

REFERENCES

Andersen, H.R. and Sørensen, H. (1985) Double low rapeseed meal in diets to young bulls. In: Advances in the Production and Utilization of Cruciferous Crops (Ed. H. Sørensen) Martinus Nijhoff Publ. Dordrecht, pp. 208-217.

Bjerg, B., Olsen, O., Rasmussen, K.W. and Sørensen, H. (1984) New principles of ion-exchange techniques suitable to sample preparation and group separation of natural products prior to liquid chromatography. J. Liquid Chromatogr. 7, 691-707.
Bjerg, B. and Sørensen, H. (1986) Isolation of intact glucosinolates bu column chromatography and determination of their purity (This volume).
Castor-Normandin, F., Gauchet, C., Prevot, A. and Sørensen, H. (1986) Le point sur l'analyse par HPLC des glucosinolate intacts de graines de colza. Revue Francaise des Corps Gras 33(3) 119-126.
Larsen, L.M. And Sørensen, H. (1985) The value of oilseed rape production in Denmark and the EEC. In: Advances in the Production and Utilization of Cruciferous Crops (H. Sørensen) Martinus Nijhoff Publ. Dordrecht, pp. 1-18.
Eggum, B.O., Olsen, O. and Sørensen, H. (1985a) Effect of glucosinolates on the nutritive value of rapeseed. ibid. pp. 50-60.
Eggum, B.O., Just, A. and Sørensen, H. (1985b) Double low rapeseed meal in diets to growing - finishing pigs. ibid. pp. 167-176.
Eggum, B.O., Larsen, L.M., Poulsen, M.H. and Sørensen, H. (1985c) Conclusions and recommendations. ibid. pp. 73-84.
Helboe, P., Olsen, O. and Sørensen, H. (1980) Separation of glucosinolates by high-performance liquid chromatography. J. Chromatography 197, 199-205.
Minchinton, I., Sang, J., Burke, D. and Truscott, R.J.W. (1982) Separation of desulfoglucosinolates by reversed-phase high-performance liquid chromatography. ibid. 247, 141-148.
Møller, P., Olsen, O., Plöger, A., Rasmussen, K.W. and Sørensen, H. (1985) Quantitative analysis of individual glucosinolates in double low rapeseed by HPLC of intact lgucosinolates. In: Advances in the Production and Utilization of Cruciferous Crops (Ed. H. Sørensen) Martinus Nijhoff Publ. Dordrecht, pp. 111-126.
Olsen, O. and Sørensen, H. (1981) Recent Advances in the analysis of glucosinolates. J. Am. Oil Chem. Soc. 58, 857-865.
Spinks, A.E., Sones, K. and Fenwick, G.R. (1984) The quantitative analysis of glucosinolates in cruciferous vegetables, oilseeds and forage crops using high performance liquid chromatography. Fette, Seifen, Anstrichm. 86, 228-231.
Sørensen, H. (1985) Limitation and possibilities of different methods suitable to quantitative analysis of glucosinolates occurring in double low rapeseed and products thereof. In: Advances in the Production and Utilization of Cruciferous Crops (Ed. H. Sørensen) Martinus Nijhoff Publ. Dordrecht, pp. 73-84.

QUICK REFLECTOMETRIC DETERMINATION OF TOTAL GLUCOSINOLATE CONTENTS IN TRADE SAMPLES

G. Röbbelen

Institute for Agronomy and Plant Breeding
University of Göttingen
von Siebold Strasse 8
D-3400 Göttingen
Federal Republic of Germany

ABSTRACT

In the framework of standardized cultivation which was performed at about 70 farms located allover the Federal Republic of Germany, three and four 00-rapeseed varieties, respectively, were produced each from a common seed source during two years, 1985 and 1986. Seed samples were taken during harvest and sent to Göttingen for analysis of their glucosinolate contents. The applied method was based on an inactivation of the endogenous myrosinase and a liberation of the glucose moiety from the glucosinolate molecule by standardized myrosinase addition. Measurement occurred by Boehringer GLUCOTEST paper and a pocket reflectometer. The results prove the method to be valid for a quick, cheap and reliable discrimination between samples derived from the traditional 0-versus the new 00-varieties of rapeseed with a high and low content of glucosinolates, respectively. The same holds true for several other analytical methods elaborated elsewhere. None of these quick tests, however, has sufficient precision and accuracy for an analysis of samples having a content of glucosinolates close to the threshold value of separation between 00- and 0-varieties. For such materials, which in fact may occur rather often in practice because of volunteer rapeseed admixtures in the fields, an accurate determination will never escape the tedious procedure of one of the more sophisticated and precise methods of quantitative glucosinolate analysis.

The developments of analytical methods for the quantitative determination of glucosinolates (GSLs) in rapeseeds have been dominated so far by the requirements of the plant breeders. Several fast and simple methods have been proposed for a first screening in large series of small seed volumes from early generations of cross progenies. A measurement of glucose release from GSLs in stoechiometric amounts at hydrolysis by the endogenous enzyme myrosinase using the GLUCOTEST paper of Boehringer was first proposed by Lein and Schön (1969) and elaborated by Lein (1970) and later Van Etten et al. (1974). Thies (1982) aiming at a higher sensitivity of glucosinolate determinations within the group of the new 00-rapeseed varieties (0-50 μmol GSL/g) defatted meal) used the formation of a yellow coloured complex of tetra chloropalladate ($PdCl_4^{--}$) with GSLs for a quick quantitative essay, since he could show that under certain conditions of analysis one Pd atom is able to combine with one GSL molecule as ligand. The absorption measured by means of a multi-channel photometer exhibited a precision sufficient for formulating a working direction for the GSL determination in large series of small volume seed samples (Thies 1983). With this palladium quicktest German plant breeders have now been working most successfully for several years reaching assessments of upto 1000 samples/day/2 persons during the seasonal peak between harvest and sowing dates in July/August (Busch and Röbbelen 1981).

Both, the glucose release and the palladium method, however, are not suited in this form for quick determination of total glucosinolate contents in larger seed lots to serve for a first discrimination of the traditional 0- from the new 00-rapeseed varieties when the farmer delivers his harvest to the silo of the agricultural trade firm. For this purpose the instrumental investment for the last method, i.e. the cost of a photometer, is too high, ranging between DM 8.000,- and DM 28.000,- . Moreover, sample processing in both cases are not adequate for large, nonuniform car-loads and they are too laborious to be conducted on the spot by the granary chief himself. Therefore, Thies (1985) had developed a new procedure based on the enzymatic glucose release from the GSLs with a following quantitative determination using a pocket reflectometer. Instruments of the latter kind had become available recently for diabetes tests and are now offered by various producers with prices between DM 500,- and 800,- only.

Working direction of the reflectometric test after Thies (1985)

Sources of experimental necessaries:

- Reflectometer "petita", serial no. A34 266: H. Wolf GmbH, Kieler Str. 33, D 5600 Wuppertal 1.
- GLUCOTEST paper, Boehringer Mannheim: (one roll sufficient for ca. 150 analyses) available in pharmacies.

- Myrosinase sets with 100 tubes of each 100 units enzyme including charcoal filterpapers, filter supports, a 25 µl Hamilton syringe etc: Bio-Chem-Color GmbH, P.O. Box 1416, D 3400 Göttingen.

Procedure

1) Grind 10 g of air dry seed in a KRUPS coffee mill for 20 seconds.
2) Loosen seed material sticking to the inner wall of the milling cavity by using a spatula and grind the total sample again for 2 seconds.
3) Weigh out an accurate amount of 5g meal using if necessary even a letter balance and
4) pour sprinkling, boiling water over this probe in a 250 ml glass beaker filling up to the 200 ml mark.
5) Stirr the substrate in the beaker by a spatula for ca.1 minute to dissolve clumps and to speed up the extraction of the GSLs.
6) Let ca.1 ml of the extract cool down to less than 60°C in a polystyrol tube of 3.5 ml volume.
7) Pour the cooled extract over into the tube containing 100 units of myrosinase. Close the tube with your thumb and mix the contents by 4 quick turns of the tube. After a reaction period of 5 minutes:
8) Add by pipetting 5 ml of a protein precipitating agent (10% chlorohexidin diacetate in methanol) and shake the tube again thoroughly by hand.
9) Draw the contents of the tube into a 1 ml syringe and press the same tightly onto the provided (see above) filter support armed with activated charcoal paper.
10) Pull out the piston from the syringe; the syringe must thereby be kept horizontal to facilitate the passage of the extract into the upper part of the filter unit.
11) Place filter unit with syringe onto a 3.5 ml polystyrol tube and collect at least 5 drops of a clear filtrate. Discard opaque or coloured filtrates resulting from an untight filter.

12) Transfer 17 µl of the filtrate onto a clean slide forming 1 cm long strip.
13) Callibrate the reflectometer with a 17 mm long piece of GLUCOTEST paper and place the same immediately afterwards onto the liquid strip on the slide. Cover with cardboard box to exclude light affects. After 1 minute repeat:
14) Callibration of the reflectometer with a second dry test-paper strip (necessary because the commercial reflectometer is programmed with 1 minute reaction time only).
15) Insert the treated testpaper with a forceps into the reflectometer after a total reaction time of 2 minutes (in the dark) and note the result of the measurement given in digital form.
16) Repeat steps 12) to 15) and calculate the mean $\bar{x}$ of both measurements.
17) Use the formula:

    $$C = (\bar{x} - 90) \cdot 0.5$$

    for calculating C, i.e. the contents of GSLs in the analysed sample expressed in µmol per g air dry seeds. The subtrahent value 90 is obtained as a mean when extracts are measured which are prepared as above but without myrosinase addition. The validity of this value should be checked, however, before using any new reflectometer instrument.

<u>Model production of 00-rapeseed</u>

During 1985 and 1986 the Ministry of Agriculture of the Federal Republic of Germany had conducted a model production of the first new 00-rapeseed varieties (seed containing less than 30 µmol GSLs/g defatted seed meal, equiv. 18 µmol GSLs/g air dry seeds). In 1985 the three varieties 'Elena', 'Liropa' and 'Lindora' and in 1986 four varieties, i.e. the above three and in addition 'Rubin', were grown in fields of at least 1 ha size in five states from Schleswig-Holstein in the north to Bavaria in the south and in 72 and 65 farms, respectively. Seed samples were taken at harvest one for each hectar field. For comparison samples were also drawn from neighbouring fields carrying traditional rapeseed with high GSL contents. All samples were mailed to Göttingen and analyzed immediately at the day of their arrival by using the reflectometric test. Later the accuracy of the measurements was checked by analyzing the same samples again by gas-liquid-chromatography of the desulfo-GSL derivatives following Thies (1979; cf. Gland et al. 1981).

The GSL content of the certified seed which was distributed to the farmers for sowing into their test fields ranged from 19.7 to 29.3 μmol/g defatted meal. In 1985 the harvest of the three 00-varieties from a total of 142 ha including 76 fields and a production volume of 420 t showed a mean of 47.2 μmol GSL/g defatted meal, but with a variation from a minimum of 180.1 μmol/g due to heavy infestation with volunteer rape plants from earlier harvests years ago. The mean GSL content of the traditional 0-varieties was 147.1 μmol/g defatted meal obtained from 110 samples provided from the same locations and farmers (Röbbelen 1986). In 1986 similar values were obtained. From a total of 146 samples of 00-rapeseed varieties a $\bar{x}$ of 24 μmol GSL/g seeds(!) and from 95 0-varieties a corresponding value of 71 μmol GSL/g was determined by accurate gas-chromatographic analysis (Figure 1).

<u>Performance of the reflectometric method</u>

Initial problems of grinding the reshly harvested seed samples by the KRUPS coffee-mill were overcome by routinely drying the incoming samples for 40 minutes at 80°C reducing the humidity of the seeds to below 8% water. The correlation of the reflectometer values with those from the final gas-chromatography was r=0.91 for the total of n=243 in 1985 and r=0.92 for n=246 samples in 1986. This high correlation indicated a high precision and accuracy of the reflectometric method; but the result is, of course, partly conditioned by the wide variation of GSL contents in the analyzed seed samples. If only those samples were considered which meet the proposed requirements for an EC intervention and subsidy, i.e. below 35 μmol/g seeds, then a correlation of only r=0.77 would

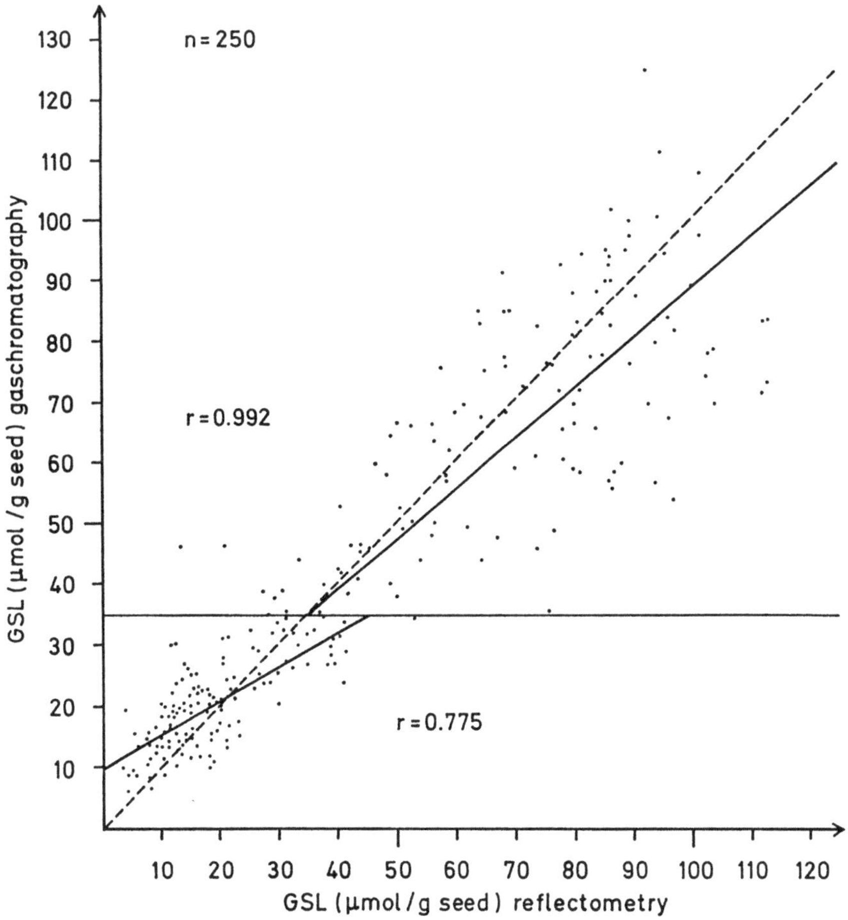

Fig. 1 Comparison of methods for the analysis of the total contents of glucosinolates (in µ mol/g seeds). 250 samples were taken from car-loads of rapeseed harvested under the "00-Rapeseed Model Production" programme, Germany F.R., 1986. GSLs were measured by the reflectometric quicktest and by liquid-gaschromatography.

apply to the 1986 data (see Figure 1). This proves that the method is sufficiently well suited to discriminate between high glucosinolate 0-rapeseed and low glucosinolate 00-rapeseed varieties; but it is really not precise enough to decide about the boarder-line cases of samples being just above or below a given threshold value.

During the whole test period a precision control chart was kept in the way that every day a glucose standard solution was measured with the used reflectometer. As seen from Figure 2 the precision of the results with a s% = 2.2 and 2.6, respectively, was indeed satisfying for a quicktest method. At August 15, however, the accuracy turned away from what had been the mean so far. The reason, to which this mistake could be ascribed was the use of a new charge of GLUCOTEST paper (charge 9NV 25A- of the same company!) with obviously different reaction norm. This case makes it very evident that a high precision of a method is not a sufficient criterion for having obtained true values. The use of an accuracy control standard is absolutely essential. In our case a glucose standard solution had been used; but a homogenous seed sample of known GSL content would serve the same purpose. In any event such standard must be analyzed routinely in certain intervals together with the candidate probes.

Searching for a useful quicktest for GSL analysis, several laboratories in Germany F.R. and England recently compared their experiences. This led to results given in Table 1. The same 8 samples of rapeseed were measured by 6 different means. The gaschromatographic method with a temperature program, prescribed by the EC Commission for their market regulations (see EWG 1986) was used as a standard for comparison. Method 1 of the quicktests was conducted as described by Heaney and Fenwick (1981) consisting of a purification of the GSL extracts by Sephadex columns, glucose release by myrosinase cleavage and colourimetry. Method 2 applied the glucotest following Lein (1970). Method 3 involved desulfation on micro-columns according to Thies (1979) and subsequent absorption photometry. Method 4 followed the prescription of the VDO (Verein Deutscher

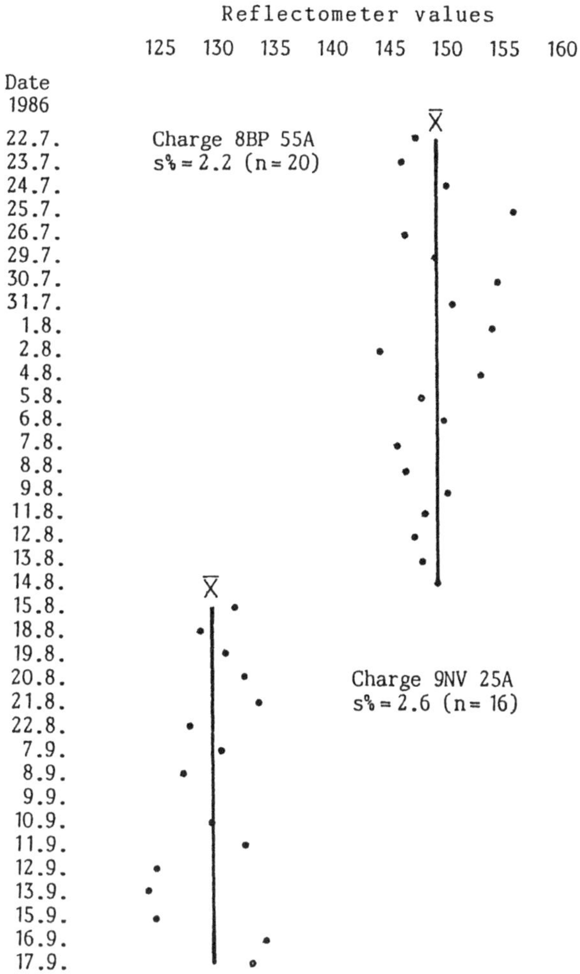

Fig. 2 Accuracy and precision control chart kept for the reflectometric GSL analysis in the Göttingen laboratory in 1986. A glucose solution Sigma 635-100 (10 μmol/ml) was daily analyzed as a standard. GLUCOTEST papers used were from charge 8BP 55A during the period July 22 - August 14 and from charge 9NV 25A during August 15 and September 17. A factor of 0.75 instead of 0.5 was used in the formula when calculating the final GSL contents in the latter series (step 17 of the working direction).

Table 1 Comparison of analytical results obtained by five different "quick methods" with those obtained by gaschromatography with temperature program (EC-method). Concentrations are given in μ mol/g seeds for 8 rapeseed samples (A-H).

Seed sample	EC method	Quick method 1	2	3	4	5
A	98	87	82	61	95	68
B	64	64	59	50	69	54
C	55	54	52	42	62	53
D	43	44	41	41	52	41
E	18	16	16	19	24	17
F	8	10	8	13	13	9
G	90	79	86	90	98	69
H	18	16	16	28	27	22

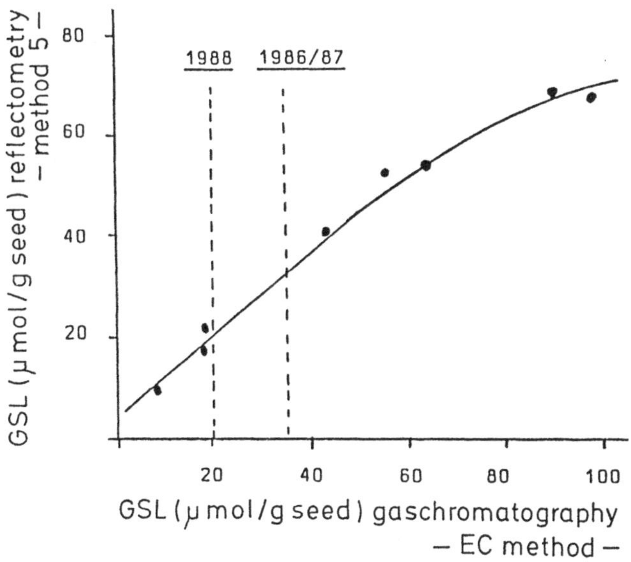

Fig. 3 Comparison of the results obtained for the above samples A-H (see Table 1) by reflectometry (method 5) and gaschromatography (EC method)

Ölmühlen) using microcolumns for desulfation as above but measurement of the desulfo GSLs by UV spectrometry. Method 5 was the reflectometric method as described above. It is easily evident from Table 1 that none of the quicktests gave results entirely out of range, again no one yielded the same values as received by the EC standard procedure. This finding is not exclusively but also explained by the fact that the EC method results in incomplete estimates of the total GSL contents too, due to at least some breakdown of indole GSLs, in particular 4OH-Glucobrassicin (see other contributions in this volume). For the reflectometric method (5) Figure 3 shows too low values for the higher GSL contents, as had been already stated by Thies (1979). But within the range set by the EC Commission as threshold values for the next two (1986/87) as well as later years (1988), the correlation between reflectometry and EC gaschromatography is sufficiently straight proving the validity of this method for quick discriminations.

Additional methods have also been proposed for quick GSL screening. Among these NIR spectometry has been investigated and a good fit was reported whenever the range of the GSL values was as wide as in Figure 1 of this paper. But the principle also holds true that the lower the GSL contents becomes through further breeding and cultivation progress, the less sensitive and precise any indirect method such as NIR will be (just opposite to the reflectometric method).

Another possibility is offered by determination of the total sulfur content of the seed using X-ray fluorescence analysis (Schnug and Haneklaus 1986). This technique indeed gives highly accurate values without tedious sample preparation in no more than 2 minutes from 5 g of homogeneously grinded meal. In both cases, however, NIR instruments and especially X-ray fluorescence spectometers mean a rather expensive investment (80.000,- to 300.000,- DM) and they are thus not applicable for the analysis of trade samples at the bin of a commercial firm where this is needed when the farmer is waiting to unload his harvested rapeseed from his truck for delivery into the high or the low glucosinolate grain silo, respectively.

Conclusion

The high precision required by the EC Commission for intervention and subsidy of 00-rapeseed lots can not be obtained by any of several quicktests for GSL determination. But each of them is almost equally well suited to discriminate between the traditional 0-rapeseed and the new low glucosinolate 00-rapeseed varieties, if these are not too highly infested by unknown amounts of volunteer rape plants. However, accuracy control standards are essential in all cases to avoid systematic errors of the analytical procedure.

ACKNOWLEDGEMENT

The author acknowledges the major contributions of Prof. W. Thies to the reported matter and the preparation of most of the presented materials. He also thanks Miss K. Fischer, Mr. U. Ammermann and Mr. R. Freter for their outstanding technical assistance.

REFERENCES

Busch, H., Röbbelen, G., 1981. Niedriger Glucosinolatgehalt als Zuchtziel für Winterraps. Angew. Bot. 55, 361-371

EWG 1986. Verordnung (EWG) Nr. 2435/86 der Kommission vom 29. Juli 1986 Anhang VIII. Raps- und Rübsensamen. Bestimmung des Glucosinolatgehalts - Amtsblatt der Europäischen Gemeinschaften Nr. L 210/56-60 v. 1.8.1986

Gland, A., Röbbelen, G., Thies, W., 1981. Variation of alkenyl glucosinolates in seeds of Brassica species. Z. Pflanzenzüchtg. 87, 96-110

Heaney, R.K., Fenwick, G.R., 1981. A micro-column method for the rapid determination of total glucosinolate content of cruciferous material. Z. Pflanzenzüchtg. 87, 89-95

Lein, K.A., 1970. Quantitative Bestimmungsmethoden für Samenglucosinolate in Brassica-Arten und ihre Anwendung in der Züchtung von glucosinolatarmem Raps. Z. Pflanzenzüchtg. 63, 137-154

Lein, K.A., Schön, W.J., 1969: Quantitative Glucosinolatbestimmung aus Halbkörnern von Brassica-Arten. Angew. Botanik 63, 87-92

Röbbelen, G., 1986. Modellanbau von 00-Raps in der Bundesrepublik Deutschland 1984/85. Raps 4, 4-10

Schnug, E., Haneklaus, S., 1986. Eine Methode zur schnellen Bestimmung des Gesamtglucosinolatgehaltes von Rapssamen. Raps 4, 128-130

Thies, W., 1979. Detection and utilization of a glucosinolate sulfohydrolase in the edible snail, Helix pomatia. Naturwiss. 66, 364-365
Thies, W., 1982. Complex-formation between glucosinolates and tetrachloropalladate (II) and its utilization in plant breeding. Fette-Seifen-Anstrichmittel 84, 338-342
Thies, W., 1983. Optimization of the "on column" desulfation and gaschromatography of glucosinolates. Proc. 6th Int. Rapeseed Conf. Paris 17-19 May 1983. Vol. II, 1343-1349
Thies, W., 1985. Glucosinolatgehalt in Handelspartien von 00-Raps - Schnelle Bestimmung durch Taschen-Reflektometer. Raps 3, 112-114
VanEtten, C.H., MacGrew, C.E., Daxenbichler. M.E., 1974. Glucosinolate determination in cruciferous seeds and meals by measurement of enzymatically released glucose. J. Agric. Food Chem. 22, 483-487

ANALYSIS OF QUALITY PARAMETERS OF WHOLE RAPESEED BY N.I.R.S.

R. Biston*, P. Dardenne*, M. Cwikowski**, J-P Wathelet** and M. Severin**

* Station de Haute Belgique, rue de Serpont, 100
6600 Libramont - Chevigny Belgium
** Faculté des Sciences Agronomiques de l'Etat,
Chaire de Chimie Générale et Organique
5800 Gembloux Belgium

ABSTRACT

Calibrations and Predictions of Nitrogen, Oil and glucosinolates content in whole rapeseed kernels were made by use of a research near infrared spectrometer (P. Sci - Mark II).

The calibration equations were developped on a large population of samples, analysed by the reference method (protein : KEJDAHL - oil : SOXHLET - glucosinolates : HPLC - GC).

Oil, protein and glucosinolates content could be estimated with high correlation and accuracy.

- For oil :
  a multiple correlation coefficient (r) of 0.98 and a standard error of estimated values (SE) of 0.53 were obtained by using 4 wavelengths after the transformation of the data (log 1/R) in second derivative .
- For protein :
  a multiple correlation coefficient (r) of 0.97 and a standard error of estimated values (SE) of 0.29 were obtained using 4 wavelengths after the transformation of the data (log 1/R) in second derivative.
- For glucosinolates :
  after the transformation of the data in first derivative, a coefficient of multiple correlation of 0.98 was observed, as well as a standard error of estimated values of 3.0 micromoles/g of seeds for samples analysed by HPLC and 3.6 micromoles/g of seeds for samples analysed by GC. The wavelengths retained were 1636 - 1624 - 1640 .

We conclude that N.I.R.S. allows an accurate prediction of the quality parameters of whole rapeseed.

INTRODUCTION

Rapid and a non destructive analytical method have a great potential for analysing cereals and oilseeds, since current methods involve seed grinding followed by slow, tedious and expensive wet chemical analytical test.

Near infrared reflectance has become used widely for the analysis of some components of a variety of plant materials especially cereals grains and oil seeds.

Near infrared has been showed to provide a satisfactory alternative to more traditional methods when dealing with the large number of samples in plant breeding programmes, in the storage and the processing plant as well as for feed manufacturers.It has been reported that N.I.R. provides a method for the simultaneous measurement of nitrogen, oil and total glucosinolate of rapeseed, yet with some practical problems during sample preparation. RIBAILLIER (1984); STARR (1985); TKACHUK (1981).

This present study describes a near infrared reflectance procedure for analysing whole rapeseed kernels. A research reflectance spectrophotometer was used to obtain N.I.R. data which were correlated with oil, protein and glucosinolate content of rapeseed samples and analysed by a stepwise multiple analyse linear regression procedure to select the wavelengths giving data which would best predict the constituents.

MATERIALS AND METHODS

Materials

Rapeseed was collected in several areas in Belgium France and Denmark during 1984-1985. Separate sets of samples were used for each of the characters studied.

After reference analyses have been carried out, sets were selected for calibration and prediction purposes. As shown in table 1, these sets had a range of the required constituent as wide as possible. For the glucosinolate a special sample selection program based on the spectra was used for calibration.

TABLE 1 Characteristics of the sets used for calibration and for prediction

Constituents	Calibration				Predictiction			
	n	Range	Mean	SD	n	Range	Mean	SD
PROTEIN (% MS)	46	21 - 26	23.8	1.12	22	22.3-25	23.9	1.22
OIL (% MS)	30	40 - 50	44.9	2.50	13	40 - 48	44.4	2.76
GLUCOSINO-LATES HPLC (MICROMOL/G)	50	4 -106	31.3	26.6	24	7 -106	23.6	20.7

Reference methods

Standard laboratory analyses were used to determine OIL. (Extraction method - Method IUPAC IB2) and protein (Kjedhal method) . The total glucosinolate content was determined respectively by HPLC (as desulfoglucosinolates - Spinks, Sones and Fenwick method) and by Gas chromatography.

The observed standard errors for the laboratory were for protein 0.1 ; for Oil 0.1; for glucosinolate 4.75 .

Near infrared method

Reflectance data were collected with a Mark II PSCi single bean spectrophotometer controlled by a North Star computer.

The data were recorded for 12 g whole rapeseed contained in a normal cylindrical rotating sample holder covered by a quartz.

For each rapeseed sample an average of two readings was recorded at each 2.0 nm interval from 1100 to 2400 nm for a total of 700 values.

The reflectance values were transformed in first or second derivative and they were linked with the laboratory values by a stepwise regression.

The highest correlation coefficient (r) and the lowest standard errors of estimate were used as criteria for evaluating the best predicting calibration equation. The equation was tested by using it to predict the same constituent in another set.

Results obtained were then compared with those determined by the standard laboratory methods.

RESULTS

The optimal wavelengths selected by the stepwise regression program for the prediction of oil, protein and glucosinolate content are shown in table 2.

TABLE 2 Optimal wavelength selection for Oil, Protein and Glucosinolate prediction in Rapeseed.

Constituants	Wavelengths
PROTEIN	1500 - 1708 - 2142 - 2164
OIL	1744 - 1712 - 1300 - 1496
GLUCOSINOLATES	1636 - 1624 - 1640

Four wavelengths are selected for prediction of protein and oil, the point 2142 - 2164 and 1744 - 1712 appear to have the most efficient effect.

For the glucosinolate determination, the selected wavelengths are similar with those published in the litterature. This spectral area seems to be very specific and highly sensitive for this constituent.

The figure 1 shows NIRS spectra (log 1/R) with 3 rapeseed samples, displaying different glucosinolate, protein and oil contents.

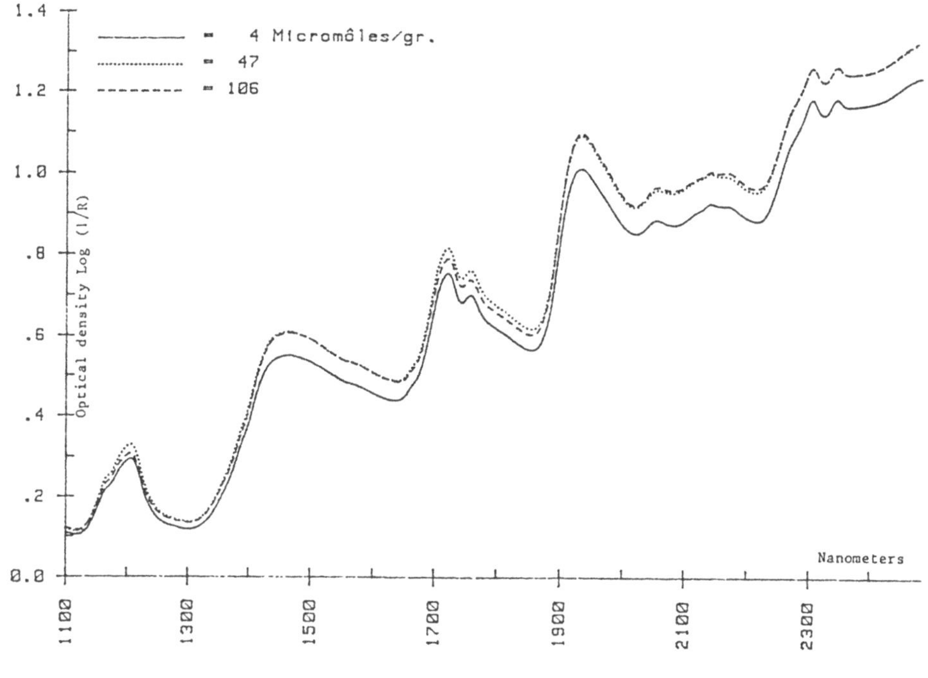

Fig. 1 NIR reflectance spectra for whole rapeseed with different content of GLUCOSINOLATE.

The difference we can observe can be explained by a granulometry effect. After a first or second derivative transformation these differences are less obvious and mainly correspond to the component variation.

For whole seeds especially it is useful to apply derivative algorithms.

Figures 2 - 4 and 6 show the derivative spectra for 3 content of each of the constituants. In each case, the spectral regions of most significant wavelengths were expanded (fig. 3 - 5 - 7).

Table 3 give the results of calibration and prediction for oil and protein.

TABLE 3 Calibration and Prediction for Protein and oil content by NIRS.

Constituents	CALIBRATION			PREDICTION			
	Math	SEC	MC	RMS	r	p	Bias
PROTEIN	D20D 20/30	.29	0.97	.32	0.97	1.03	0.18
OIL	D20D 20/0	.51	0.98	0.58	0.98	1.00	0.07

Table 4 give the results of calibrations and predictions for glucosinolate.

TABLE 4 Calibrations and Predictions for Glucosinolate content by NIRS

Constituents	CALIBRATION			PREDICTION			
	Math	SEC	MC	RMS	r	p	Bias
Total Desulfo-glucosinolate HPLC	D10D 10/0	3.08	0.99	2.96	0.99	0.97	0.55
Total Desulfo-glucosinolate G.C.	D10D 10/0	3.39	0.99	3.60	0.98	0.97	0.09
Alkenyl-glucosinolates HPLC	D10D 10/0	2.60	0.99	3.21	0.99	0.98	-.895

The figures 8 to 11 give the calibration plot for glucosinolate content (HPLC and G.C method) v.s. NIR reflectance.

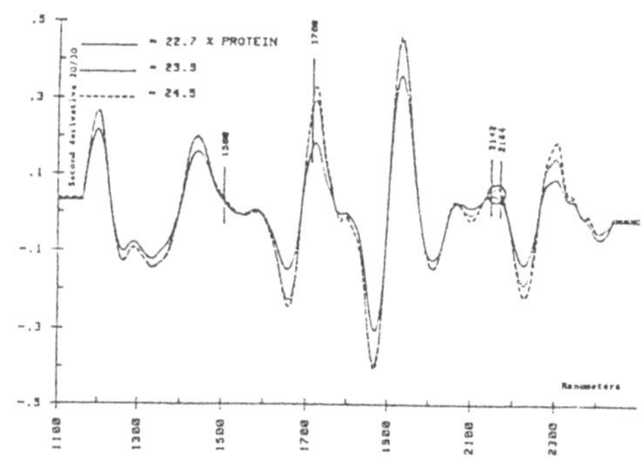

Fig. 2 NIR second derivative spectra for whole rapeseed with different content of PROTEIN

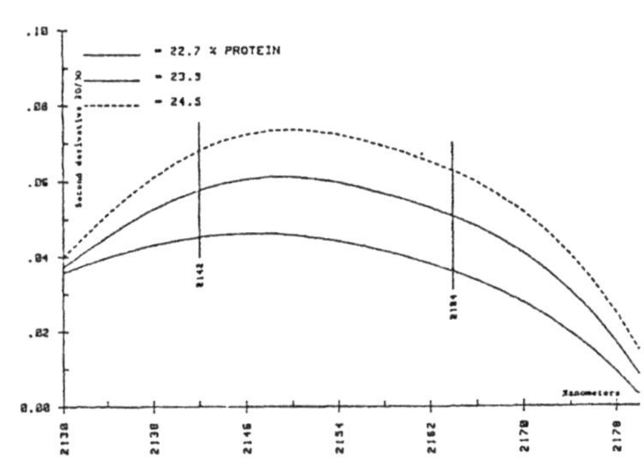

Fig. 3 NIR second derivative spectra for whole rapeseed with different content of PROTEIN

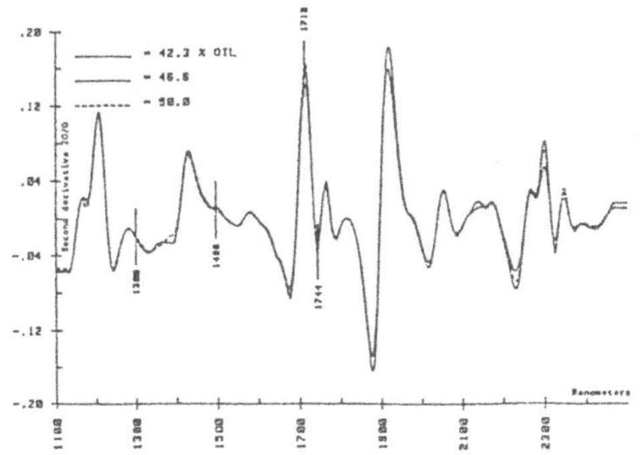

Fig. 4 NIR second derivative spectra for whole rapeseed with different content of OIL

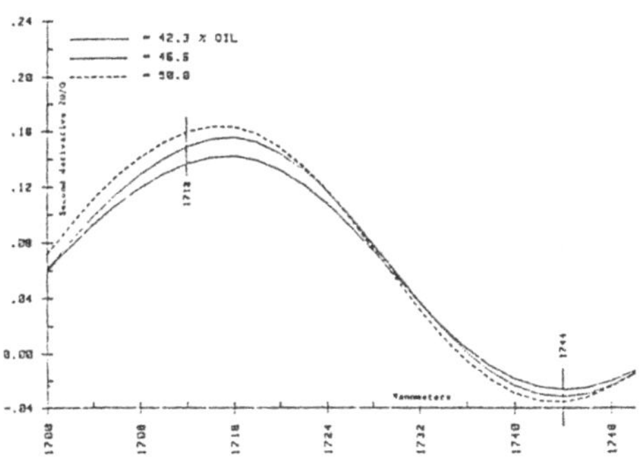

Fig. 5 NIR second derivative spectra for whole rapeseed with different content of OIL

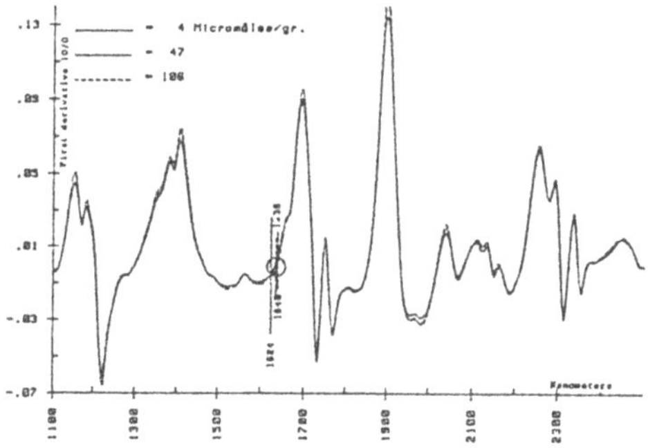

Fig. 6 NIR first derivative spectra for whole rapeseed with different content of GLUCOSINOLATE

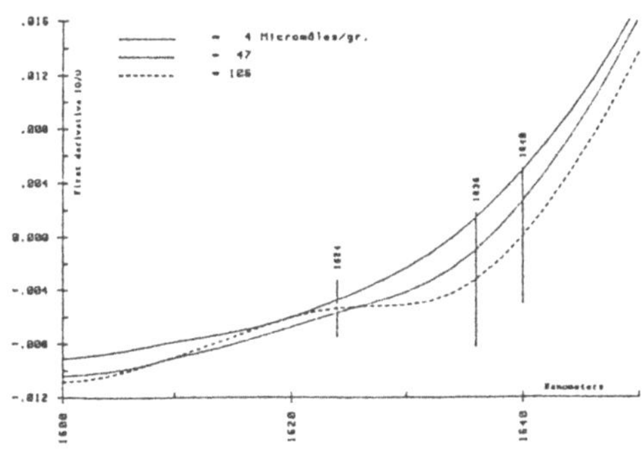

Fig. 7 NIR first derivative spectra for whole rapeseed with différent content of GLUCOSINOLATE

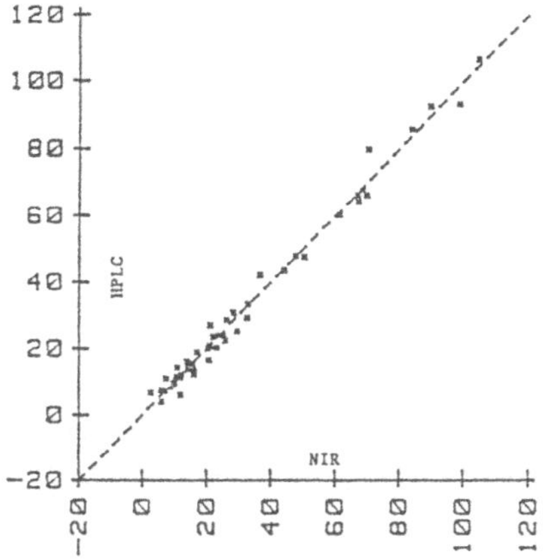

Fig. 8 Calibration plot for GLUCOSINOLATE content HPLC method vs NIR reflectance.

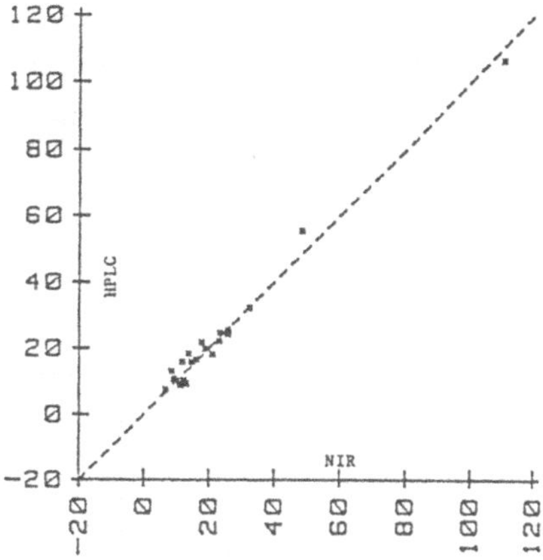

Fig. 9 Prediction plot for GLUCOSINOLATE content HPLC method vs NIR reflectance.

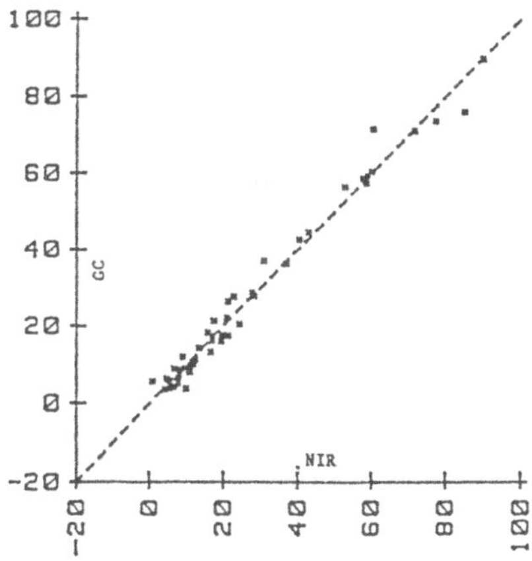

Fig. 10 Calibration plot for GLUCOSINOLATE content GC method vs NIR reflectance

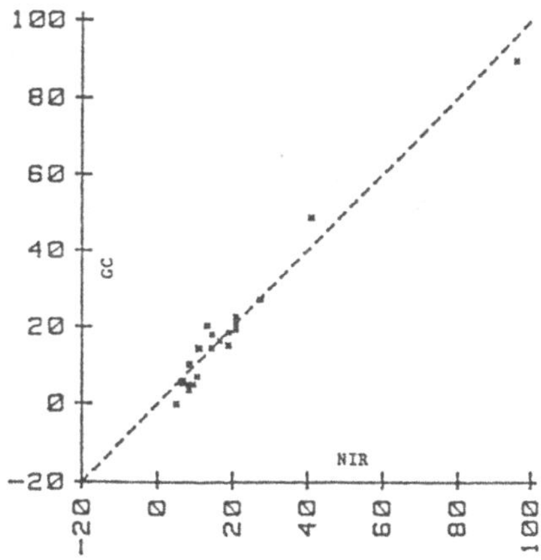

Fig. 11 Prediction plot for GLUCOSINOLATE content GC method vs NIR reflectance

CONCLUSIONS

The results obtained by using a scanning spectrophotometer indicate that NIR is a suitable method for the prediction of oil, protein and glucosinolate content in whole rapeseed.

An acceptable level of accuracy can be obtained by using a second derivative of the data, which minimize the particle size variation on the factors affecting the general level of absorbance.

Specific wavelengths were obtained to predict the glucosinolate content with a high degree of accuracy.

The speed for analysis (1 minute for 3 constituents), the non destructivess of the seed make this technique well adapted for breeding purpose as well as for quality control in oil factories and in feed manufactures.

REFERENCES

Ribaillier,D.,1984. L'analyse des graines oleagineuses par spectroscopie de reflexion dans le proche infra rouge. Revue Française des Corps gras, 4-5 .

Spinks, E., Sones, K. and Fenwick, G.R. ,1984. The quantitative analysis of glucosinolates in cruciferous vegetables, oilseeds and forage crops using high performance liquid chromatography. Fette, Seifen, Anstrichmittel, 86 , 228-231.

Starr, C., Suttle, J., Morgan, A. and Simth, D., 1985. A comparison of sample preparation and calibration techniques for the estimation of nitrogen, oil and glucosinolate content of rapeseed by near infrared spectroscopy. J. Agric. Sc. Camb. 104 , 317-323.

Tkachuk, R., 1981. Oil and protein analysis of whole rapeseed kernels by near infrared reflectance spectroscopy. Journal of the American Oil Chemists Society, 58, 819 - 822.

GLUCOSINOLATE ANALYSIS IN WHOLE RAPESEED BY NEAR INFRARED REFLECTANCE SPECTROSCOPY

M. RENARD*, C. BERNARD**, M. DESCHAMPS*, V. FURTOSS***, M. LILA***, A. QUINSAC****, J.M. REGNIER** and D. RIBAILLIER****

* INRA - STATION D'AMELIORATION DES PLANTES - F.35650 LE RHEU
** UCAAB - F.02400 CHATEAU THIERRY
*** INRA - STATION D'AMELIORATION DES PLANTES FOURRAGERES - F.86600 LUSIGNAN
**** CETIOM - LABORATOIRE D'ANALYSES - F.45160 OLIVET

ABSTRACT

A monochrometer-type near infrared spectrophotometer was used to predict glucosinolate content in whole rapeseed (B. napus L.). Calibration equations were established by multiple linear regression of trimethylsilylated desulphoglucosinolates evaluated by GLC on NIRS data from different populations. A coefficient of multiple determination (R^2) of 0.95 was obtained. In most cases standard errors were not higher than commonly reported for conventional manual assays. Then calibration was studied with a filter-type near infrared photometer equipped with 19 filters. Using 6 wavelengths (1445, 1620, 1632, 1640, 2139, and 2208), the coefficient of determination was 0.94 and standard error slightly higher.

INTRODUCTION

Improving nutritional quality is an important goal in breeding rapeseed (B. napus L.) varieties. High concentrations of glucosinolates in seeds are one of the major problems limiting the utilization of this crop, intact glucosinolates and the degradation products causing various antinutritional effects. Glucosinolate content in rapeseed has been assayed by various procedures (SORENSEN, 1985). However, all these methods involve seed grinding.

Near infrared reflectance spectroscopy (NIRS) has been investigated as a method of accurate and rapid quality prediction (CLARK, 1985). Thus, NIRS has been increasingly used for the determination of protein, oil and moisture content in grains and oilseeds. For protein and oil of whole rapeseed kernels, a high correlation has been shown between analyses by standard laboratory methods and estimates by NIRS (TKACHUK, 1981).

The objective of this study was to evaluate NIRS for a rapid and accurate prediction of glucosinolates in whole rapeseed with regard to application to plant breeding and oilseed rape production.

MATERIALS AND METHODS

Three populations were investigated. Samples were harvested from experimental plots in FRANCE (populations 1 and 3) and in SWEDEN (population 2). A standard analytical procedure as described by THIES (1976, 1979) was used to determine the glucosinolates by gas liquid chromatrography of the trimethylsilylated desulphoglucosinolates (table 1).

TABLE 1 Glucosinolate range and mean of each rapeseed population (μ moles/g of defatted meal).

	Number of samples	Range	Mean
Population 1	54	6.8-220.5	99.2
Population 2	30	18.1-122.3	57.1
Population 3	20	5.3-195.3	87.9

In this study calibration equations were obtained with two different instruments : a monochrometer-type near infrared spectrophotometer and then a filter-type near infrared spectrophotometer. The monochrometer-type instrument was a technicon Infraalyser 500 interfaced with a HP 1000-A600 computer. The instrument can scan the near infrared region from 560 to 2650 nm. Data were recorded as Log 1/R where R = reflectance. Calibration equations were calculated by multiple linear regression of the manual values on the Log 1/R values using two procedures : a stepwise regression program (STEP-UP) and the COMBO program which permits scanning of all the combinations of n wavelengths among 100 ($n<6$).

The filter-type instrument was a technicon model Infraalyser 400R containing 19 interference filters mounted on the circumference of a rotating wheel. The instrument was fitted with filters at the following wavelengths (nm) : 1445, 1620, 1632, 1640, 1680, 1722, 1734, 1759, 1778, 1818, 1940, 1982, 2100, 2139, 2180, 2190, 2208, 2230, 2310, subsequently referred to as filters 2 to 20. Filters 3, 4, and 5 were manufactured after the study on the monochrometer-type near infrared spectrophotometer.

RESULTS AND DISCUSSION

Monochrometer-type near infrared spectrophotometer :

Using population 1 and the COMBO program with a step of 14 nm, the best combination of 5 wavelengths was established with wavelengths between 1600 and 1680 nm. In this range, the choice was improved with a step of 4 nm. The 5 selected wavelengths are the following :

Equation 1 : 1620, 1632, 1650, 1660, 1674.

Searching for a combination using some of the standard wavelengths of the filter-type instrument, two other equations were obtained :

Equation 2 : 1620, 1632, 1640, 2208, 2230 (COMBO program).

Equation 3 : 1445, 1620, 1632, 1640 (STEP-UP program).

Then these equations were applied to the different populations. The coefficients of multiple determination (R2) were between 0.93 and 0.97, standard errors (SE) between 7.8 and 16.0 (table 2), in most cases not higher than commonly reported for conventional rapid manual assays.

TABLE 2 Calibration data obtained on three populations of whole rapeseed with a monochrometer-type instrument.

		Population 1	Population 2	Population 3
Equation 1				
	R^2	0.96	0.94	0.96
	SE	12.4	7.8	12.2
Equation 2				
	R^2	0.97	0.92	0.97
	SE	13.7	9.8	12.7
Equation 3				
	R^2	0.93	0.95	0.95
	SE	16.0	8.0	15.2

- Filter-type near infrared spectrophotometer :

With population 3 and equations 2 and 3, the coefficients of determination were lower than the coefficients determined with the monochrometer-type instrument and the standard errors slightly higher (table 3). By multiple linear regression, the best combinations of n filters were respectively (table 3) for :

- 3 filters : 1620, 1632, 1640 (equation 4),
- 4 filters : 1445, 1620, 1632, 1640 (equation 3),
- 5 filters : 1445, 1620, 1632, 1640, 1982 (equation 5),
- 6 filters : 1445, 1620, 1632, 1640, 2139, 2208 (equation 6).

TABLE 3 Calibration data obtained on population 3 with a filter-type instrument.

Equations	2	3	4	5	6
Filters	3.4.5 18.19	2.3 4.5	3.4.5	2.3.4 5.13	2.3.4.5 14.18
R2	0.92	0.93	0.91	0.93	0.94
SE	16.7	15.8	17.6	15.7	14.3

In conclusion, our study confirms that the optimal wavelengths for glucosinolates are between 1620 and 1674 nm (TKACHUK, 1981). With reasonable care, the monochrometer system provides an acceptable means of analysing rapeseed for glucosinolates. But the filter-type instrument needs further investigations before direct application to plant breeding and rapeseed production.

REFERENCES

CLARK, D.H. 1985. History of NIRS analysis of agricultural products. U.S. Dept. Agric. Agric. Hondboock., n° 643, 7-11.

SORENSEN, H. 1985. Limitations and possibilities of different methods suitable to quantitative analysis of glucosinolates occurring in double low rapeseed and products thereof in "world crops : production, utilization, description vol 11 : advances in the production and utilization of cruciferous crops" (Ed. H. SORENSEN) (MARTINUS NIJHOFF, W. JUNK Publishers) pp. 73-84.

THIES, W. 1976. Quantitative gas liquid chromatography of glucosinolates in a microliter scale. Fette. Seifen. Anstrichm., 78, 231-234.

THIES, W. 1979. Detection and utilization of a glucosinolate sulfohydrolase in the edible snail, Helix pomatia. Naturwissenschaften. 66, 364-365.

TKACHUK, R. 1981. Oil and protein analysis of whole rapeseed Kernels by near infrared reflectance spectroscopy. J.A.O.C.S. 58, 819-822.

FUTURE REQUIREMENTS FOR GLUCOSINOLATE ANALYSIS - A U.K. VIEW.

R.K. Heaney and G.R. Fenwick

Bioactive Components Group, AFRC Institute of Food Research
(Norwich Laboratory), Colney Lane, Norwich, NR4 7UA, U.K.

RACT

Progress towards the development of an official high performance id chromatographic method for glucosinolate analysis is reviewed and against recent advances in this area. Emphasis is placed on the nization of various stages in the analytical procedure and their rporation into the final method. The problem of the availability of internal standard glucotropaeolin, the concern of U.K. breeders and hers over the lack of suitable methods for monitoring seed and the for a reliable rapid method are stressed. Work in these areas at the itute of Food Research (Norwich) is briefly mentioned. The lopment of techniques for monitoring the effect on glucosinolates of seed "detoxification" processes (which in turn are taken as an index utritional quality) and for analysing the metabolic breakdown products lucosinolates and their conjugates in animals fed rapeseed meal (and ns consuming brassica vegetables) are suggested as priority areas for re research.

:ODUCTION

The purpose of this paper is to present the priorities for future :osinolate research as perceived from the UK. The importance of :osinolates (and thereby glucosinolate analysis) to the whole rapeseed ıstry, comprising breeders, growers, crushers, animal feed compounders livestock farmers, is obvious. Within this industry, 'glucosinolate' an increasingly emotive word and glucosinolate analysis a confusing -if confused- area. Against such a background international meetings of erts are valuable both to the delegates, who can meet and discuss new hniques, modifications of existing methods and - undoubtedly - the tinuing problems, and to the individual national industries who are reby informed of recent advances and forthcoming developments.

pite unfavourable climatic conditions in the winter of 1985-6 which led the ploughing in of a significant minority of the area under winter eseed cultivation (some of which was thereafter drilled with spring eseed), over 850,000t of rapeseed is expected to be gathered in the rent harvest, approximately 25% of the total EC figure. The UK crop is almost exclusively 'high' glucosinolate in type, but in the face of EC es intended to support the switch to 'intermediate' or 'low' varieties significant changes will occur over the next 5 years. It is against this background, and because rapeseed meal is utilised in the UK almost exclusively in cattle rations, rather than as a replacement for imported soya products in pig and poultry rations, that the authors' comments should be considered.

In addition, given the high intake of inexpensive cruciferous vegetables in the UK and the trend, supported by the medical and nutritional professions, toward even higher consumption of such leafy green vegetables, there is currently concern about the levels and effects of dietary glucosinolates in man. This concern is shared by workers in other countries and it is interesting to note the paper of Dr. Muuse and colleagues describing work in this area in the Netherlands. Aspects of such work in the UK will be also highlighted.

Finally since the question of (individual) glucosinolate content and analysis is arguably most important in the context of animal feeding comments will be made on the importance of establishing user-confidence in rapeseed meal. It is suggested that this can only be done in realistic (and necessarily, expensive) feeding trials - these will also establish the analytical requirements for the future, both for glucosinolates and other classes of endogenous toxicant.

ANALYSIS

During the last 10 years the development of methods for the analysis of glucosinolates in rapeseed has followed two clearly separate routes;

(i) methods designed to accurately quantitate the individual glucosinolates.

(ii) methods which give a measure only of total glucosinolates.

Early gas chromatographic (g.c.) methods which depended on the analysis of volatile breakdown products after hydrolysis with the enzyme myrosinase, were clearly inaccurate since it was necessary to separately quantitate by U.V. absorption methods progoitrin and gluconapoleiferin, both important constituents of single zero rapeseed.

Individual glucosinolates by chromatographic methods

TABLE 1 Some stages in the development of chromatographic methods for the analysis of individual glucosinolates

	g.c. methods	h.p.l.c. methods
Breakdown products	Youngs and Wetter (1967)	Maheshwari *et al.* (1979)
Intact glucosinolates	Underhill and Kirkland (1971) Thies (1976)	Helboe *et al.* (1980) Møller *et al.* (1984)
Desulphoglucosinolates	Thies (1977), (1979) Heaney and Fenwick (1980a) Daun and McGregor (1983)	Minchinton *et al.* (1982) Spinks *et al.* (1984) Sang and Truscott (1984)

Such methods, which are still retained as official British Standard and International Standards Organisation methods, also neglect the indole glucosinolates (glucobrassicins) which constitute a high proportion of the glucosinolates in double zero rapeseed and led to the development (Thies 1976, 1977, 1979) of more sophisticated methods wherein glucosinolates were first desulphated and then derivatised before separation using isothermal g.c. The application of temperature programming to this method (Heaney and Fenwick 1980a) made possible the analysis of glucobrassicin and 4-hydroxyglucobrassicin, compounds which were to assume greater significance with the advent of double zero rapeseed. This approach, more recently described by Daun & McGregor (1983), was proposed by the E.C. as an interim official method (E.E.C. 1985) and it is also currently being examined by the International Standards Organisation.

The suitability of high performance liquid chromatography (h.p.l.c.) for the separation of intact glucosinolates (Helboe *et al.*, 1980) or of desulphoglucosinolates (Minchinton et al., 1982) prompted further studies aimed at putting this highly promising technique onto a fully quantitative basis. However, the basically different approaches of the earlier methods (reviewed by McGregor et al., 1983) continued to be followed and the quantitative analysis of intact glucosinolates by h.p.l.c. was described Møller et al (1984) at about the same time as methods based on desulphoglucosinolates were published by Spinks et al. (1984) and Sang and Truscott (1984). These developments led the EEC Committee of Experts on the Analysis of Glucosinolates in Rapeseed to propose further examination of these techniques with a view to agreement on the establishment of an analytical protocol. The suggested time scale for this study was originally 2-3 years - which may in retrospect be seen as an underestimate - and since that time a considerable amount of effort has been expended in different laboratories in attempts to make the h.p.l.c. approach truly quantitative.

During this period, meetings in the U.K. involving national and international agencies, seed breeders, seed crushers, feed compounders and analysts have identified many problems - some of which (methods of extraction, choice of internal standard, response factors etc.) are common to both g.c. and h.p.l.c. procedures. Whilst it was evident that research groups were adopting a broadly similar approach to the analysis of glucosinolates using h.p.l.c., there were considerable differences in detail. For example, a confusing variety of extraction techniques are in current use. At a recent UK meeting, representatives from several countries described the use of water, aqueous ethanol, aqueous methanol and acidified aqueous methanol. It was not clear whether clarification with barium/lead reagent was necessary and one laboratory reported significant losses of both gluconapin and 4-hydroxyglucobrassicin when this reagent was used. Similarly, although it has been claimed that the use of mercaptoethanol improves the yield of 4-hydroxyglucobrassicin, the reagent has not been adopted universally.

The enzymic desulphation of glucosinolates prior to g.c. or h.p.l.c. analysis effects a highly specific clean-up step and is in use in many laboratories in Europe and elsewhere. The technique is however open to wide interpretation with different concentrations of enzyme, pH, time and temperature all finding current use. It has been reported that glucosinolates can occur bound to sinapic, and other, acids (Sørensen, 1985) and these compounds have been reported at levels of 5 μmole g^{-1} in some double zero rapeseed samples analysed in France. The occurrence in rapeseed of these 'bound' glucosinolates, their stability toward processing and their behaviour under various analytical conditions clearly require urgent study. The first requirements however are the isolation and chemical characterisation of these compounds together with an assessment of their stability over a range of model conditions.

Because sinigrin is a major constituent in the seed of some cruciferous weeds likely to occur as contaminants of rapeseed, it is now generally accepted that this compound, although easily available, is not an appropriate internal standard. McGregor (1985) has suggested the use of O-nitrophenyl-β-D-galactopyranoside but the consensus of present opinion supports the view that glucotropaeolin (benzyl glucosinolate) should be used. However this compound is no longer available commercially in Europe and accordingly an approach has been made to a major chemical company. The potential of the market is currently being assessed and it is hoped that pure glucotropaeolin will soon be available.

The lack of reliable response factors remains a major obstacle to the quantitative analysis of glucosinolates. Values have been published (by McGregor, 1985; Spinks et al., 1984; Sang and Truscott, 1984 amongst others) but a proper comparison is difficult at present due to the use of different solvents, different internal standards and most important, different detector wavelengths - the choice of the latter sometimes being determined by the availability of suitable filters. Such a comparison, ideally carried on in a number of laboratories is a matter of some urgency. It is encouraging that many of these problems are now being addressed with the results of some of these studies being presented at this meeting. Inspite of the many permutations possible at each stage of the analysis, attempts have been made within the framework of the EC to make comparisons between the method of Møller et al. (1984) based on h.p.l.c. with ion-pairing of intact glucosinolates and that of Spinks et al. (1984) in which the glucosinolates are initially desulphated. Each method has its merits but it is essential that a degree of uniformity be achieved in the areas mentioned above, before either h.p.l.c. procedure can be satisfactorily judged.

Given the quantity and quality of scientific expertise being applied to this area it is to be regretted that more progress toward a unified method has not yet been made. It would seem to be important, once a particular aspect of the analytical technique has been optimised following thorough scientific study, that this should be recommended throughout the Community for inclusion into the final protocol. The method should however be subject to periodic review in order to take account of new developments (the comparitively recent discoveries of 4-hydroxyglucobrassicin and of the 'bound' glucosinolates illustrates this latter point). Without such a coordinated and integrated approach, progress toward the common goal will remain slow.

In the U.K. the question of response factors has been addressed so far with only limited success, by isolation of intact glucosinolates and, indirectly, by the method of McGregor (1985). The glucosinolates of rapeseed have so far proved difficult to isolate in a pure form and although the method of McGregor (1985) offers great promise of an early solution to this problem, rather poor repeatability and non-specific colour formation have hindered progress. The problem of the reliable identification of desulphoglucosinolate peaks using h.p.l.c. has been greatly aided by the successful on-line coupling of mass spectrometry to h.p.l.c. column output (Mellon and Spinks, 1986). Using the thermospray LC/MS technique it is now possible to assign and positively identify (by means of protonated molecular ions and characteristic fragment ions) all of the major desulphoglucosinolates in an h.p.l.c. analysis of cruciferous seed or leaf material. This technique in combination with an accurate assignment of response factors using the method of McGregor (1985) would allow the quantitation of many lesser known glucosinolates. Although LC/MS would find little application in rapeseed analysis in which the main glucosinolates are well known, it has recently been used to confirm the identity of unknown peaks in the h.p.l.c. analysis of calabrese varieties (Lewis and Fenwick, 1986).

Total glucosinolates - rapid methods

Although the development of a definitive method remains an important objective there is increasing concern that such a method falls short of that required by plant breeders and seed crushers in two important respects - speed and cost.

The recent introduction by the EEC of premium payments for rapeseed with a glucosinolate content of less than 60μmoles per gramme of seed and the eventual (in 1990-91) withdrawal of support for material in excess of this (or a suitably tapered) figure has lent a new sense of urgency to the development of rapid methods for glucosinolate analysis.

U.K. crushing mills are generally supplied by individual 5-20 tonne lorry loads, each requiring immediate analysis in order that the load may be directed into the correcting holding bin. This problem will be particularly acute in the U.K. during the changeover from high to low glucosinolate rapeseed due to the mixed nature of deliveries and since it is probable that in the U.K. some single low seed will still be produced in 1991-92. The problem is exacerbated by the fact that most U.K. seed is delivered direct from the farm via a trader or merchant rather than in larger loads combining cooperative or contract loads. A rapid turn round of lorries is therefore important since the crusher is charged for excess stand-by time at the plant. A maximum time of 10 minutes has been suggested and this obviously eliminates methods based on g.c. or h.p.l.c. Such sophisticated methods would also be expensive to operate routinely, thus eroding further the financial benefit gained from the small premium paid to compensate for the reduced yields with some low glucosinolate varieties. For routine screening purposes analytical methods should be inexpensive to apply and should be capable of being conducted and interpreted by a laboratory technician or chemical assistant.

Methods for the analysis of total glucosinolates are considerably faster than methods for individual glucosinolates and are usually amenable to several samples being analysed simultaneously. With such techniques extraction of the glucosinolates is always a necessary first step. Hydrolysis of the extracted glucosinolates using the enzyme myrosinase followed by determination of the released glucose forms the basis of a number of methods. Such methods, involving a preliminary ion exchange step to remove endogenous glucose and other interfering compounds have been described by Lein (1970), VanEtten et al. (1974) and Heaney and Fenwick (1981). Other workers have further reduced these time-consuming procedures by the use of (Craig and Draper, 1979) and by simplifying the colorimetric analysis by use of Glucose Test paper or a pocket reflectometer (Thies, 1985). A method which attracted considerable interest at the 1983 Rapeseed Congress in Paris (Thies, 1982) has since been further developed by Møller et al. (1985). This depends on the ability of extracted intact glucosinolates to form complexes with palladium salts. The method compares well with other methods and has been used in a number of laboratories.

It seems probable however that the only methods which are likely to meet the 10 minute time limit required by the crushing mills will be those in which even the extraction step has been eliminated. In order to achieve this target it may be necessary to sacrifice accuracy and possibly, specificity and accept a method capable of distinguishing clearly between rapeseed with glucosinolates in excess of 70 μmole/g (rejected) and rapeseed with less than 50 μmole/g (accepted). Material with levels between these two values would be analysed using a more accurate procedure. In passing the possibility of using the enzymically-released sulphate as a measure of total glucosinolate content should not be ignored, although at present existing methods fall far short of the requirements, both in speed and sample throughput.

Methods based on the use of near-infrared reflectance spectroscopy (N.I.R.S.) have been published (Starr et al. 1985, Lila et al, 1986) and the potential of X-ray fluorescence spectrometry is being examined - such methods would be fast and simple to operate. In particular the widespread industrial use of N.I.R. spectrometers makes this a particularly attractive technique and in preliminary work at the Institute of Food Research (Norwich) samples of whole rapeseed of known glucosinolate content were scanned by N.I.R.S. at 24°C using a Neotec Mark 1 scanning instrument (Pacific Scientific). The data was subsequently transferred to a DG10 computer (Data General) and analysed using SPICE and CSAS software.

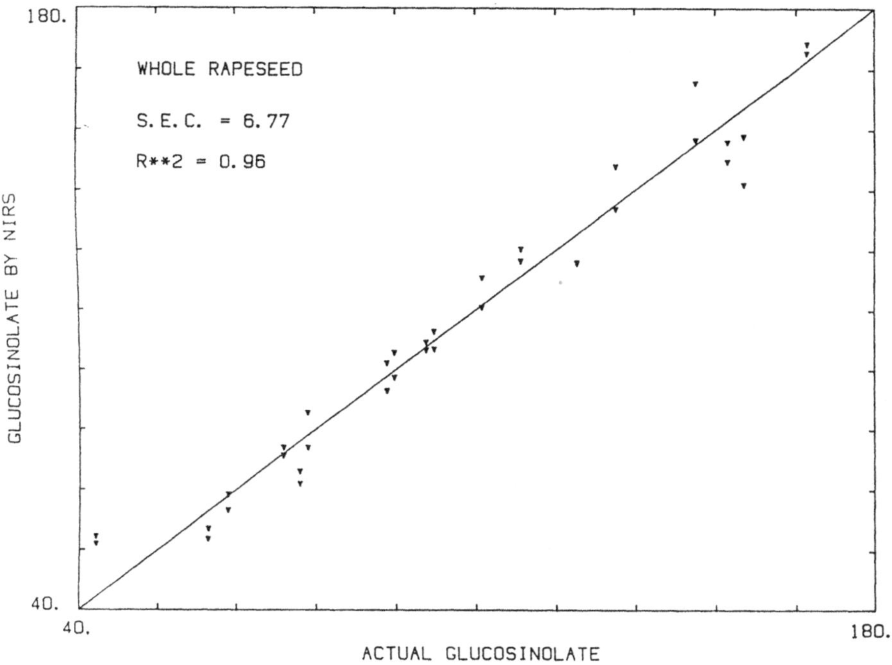

FIGURE 1. Calibration graph for the analysis of glucosinolate content in whole rapeseed.

The results from the initial regression analysis were encouraging (Figure 1). A four wavelength calibration equation gave a standard error of calibration of 6.77 and a correlation coefficient of 0.96. Future work, supported by the Ministry of Agriculture, Fisheries and Food, will involve the analysis of larger and more varied data sets and investigate the relevance of the chosen wavelengths.

NUTRITIONAL AND BIOLOGICAL IMPLICATIONS

Although much of the interest in glucosinolates is related to the antinutritional effects in animals fed rapeseed meal, there is also concern about the levels of these compounds in cruciferous crops consumed by man. Recent studies of glucosinolate levels in such crops are summarised in Table 2.

Table 2. Some recent studies of glucosinolate content of cruciferous crops

N. America	
Cabbage	VanEtten *et al*. 1976
Chinese cabbage	Daxenbichler *et al*. 1979
Turnip	Carlson *et al*. 1982
Rutabaga	
United Kingdom	
Brussels sprouts	Heaney and Fenwick 1980b
Cabbage	
Swede	Sones *et al*. 1984a
Turnip	
Cauliflower	Sones *et al*. 1984b
Fodder brassicas	Bradshaw *et al*. 1984
Calabrese	Lewis and Fenwick 1986

The levels of glucosinolates in brassicas consumed by man is of particular interest in view of the suggestion that brassica vegetables exert a protective effect against cancer. This effect has been shown to be attributable, in part at least, to breakdown products of glucobrassicin (Wattenberg and Loub, 1978). Furthermore, these compounds have been demonstrated to be capable of inducing hepatic mixed function oxidase activity *in vivo* (McDannell *et al*., 1986). However, a recent study (Wakabayashi *et al*. 1985) has shown that indole-3-acetonitrile can react with nitrite to form compounds which exhibit *in vitro* mutagenicity. This has also been demonstrated to be the case with 5-vinyloxazolidine-2-thione (Lüthy *et al*. 1984). The significance of these findings in man has, however, yet to be ascertained. 5-Vinyloxazolidine-2-thione, produced enzymically from the ubiquitous glucosinolate progoitrin, is also a potent goitrogen and inhibitor of hepatic enzyme systems. A recent study with human volunteers fed a brassica diet rich in progoitrin, showed no effect on thyroid hormone levels (McMillan *et al*. 1986), possibly because of inactivation of myrosinase during cooking. Currently the effect of this compound and brassica vegetables on hepatic trimethylamineoxidase inhibition (trimethylaminuria) is under clinical investigation.

Up to 20% rapeseed meal may be included in some ruminant rations, but this valuable protein source finds little use in poultry or monogastric rations due to its content of antinutritional compounds foremost of which are glucosinolates. Processes to reduce or remove glucosinolates (often referred to as "detoxication") include extrusion or micronization with or without added chemicals and such treatments have been monitored by observing the reduction of glucosinolates and the formation of oxazolidinethiones and isothiocyanates. Under certain conditions (Fenwick et al., 1986) large amounts of hydroxynitriles may result and the toxic nature of these compounds would suggest that a reduction in glucosinolate levels does not necessarily equate with detoxication. The presence of such breakdown products does not quantitatively explain the loss of glucosinolates and it is important that the nature and amounts of all of the breakdown products be established. A research programme supported by governmental and industrial funding is currently being undertaken in the U.K. to estimate the breakdown products present in rapeseed meal processed in a variety of ways and in animal tissues and in food (milk and eggs).

Similarly, although there have been many studies on the nutritional effects of the use of rapeseed meal in animal diets, very little is known about the subsequent fate of glucosinolates in animals (or in man consuming brassica vegetables). Studies have been conducted on the effects in animals of individual intact glucosinolates (Bille et al., 1983, Vermorel et al., 1986) but currently little is known about their metabolism, conjugation or detoxification. An understanding of such processes and their biological consequences in individual livestock species of commercial importance (especially pigs, poultry) is a prerequisite if the full potential of rapeseed meal is to be realised. Livestock farmers will only use rapeseed meal in rations if there is confidence that it will not cause problems. Equally the opportunity of using rapeseed meal will only arise if the compounder or feedstuff manufacturer has confidence in the product. This confidence can only be established by rigorous - and costly - feeding trials in which the results are evaluated in terms of both their statistical - and financial -

significance. Such trials require organisation and, in view of their cost, every effort should be made to maximise the usefulness of the data, or availability of samples, e.g. conducting rigorous sensory evaluation on any meat, poultry and dairy products. Such studies are beneficial internationally, rather than nationally, and might usefully be funded or supported on an international cooperative basis. It is to be hoped that the EEC will in the future, as in the past, play a significant part in such work.

Rapeseed offers a very clear example of the mutuality of industrial interest - the breeders, farmers, crushers, compounders and livestock rearers being united by the objective of producing a rapeseed which can be widely fed to livestock and thereby replace costly imports. This objective also offers an opportunity for the blending of basic science, industrial technology and commercial exploitation. That rapeseed analysis - and more specifically the analysis of rapeseed for glucosinolates - is vital to the future, is clear. If there are exciting times ahead for the industry, that is also true for the analyst.

ACKNOWLEDGEMENT

Some of the work referred to in this paper is funded by the Ministry of Agriculture, Fisheries and Food. The authors gratefully acknowledge the advice and assistance of colleagues in plant breeding and in the rapeseed industry especially Dr. A. Morgan (NIAB) and Dr. R. Mawson (Unilever). Dr. F. Mellon and Mr. A. Grant of this Institute are thanked for permission to describe preliminary work which will be published in full elsewhere.

REFERENCES

Bille, N., Eggum, B.O., Jacobsen, I., Olsen, O. and Sørensen, H. 1983. Antinutritional and toxic effects in rats of individual glucosinolates (± myrosinases) added to a standard diet. Effects on protein utilization and organ weights. Z. Tierphysiol. Tierernaehr. Futtermittelk, 49, 195-210.

Bradshaw, J.E., Heaney, R.K., MacFarlane-Smith, W.H., Gowers, S., Gemmell, D.J. and Fenwick, G.R. 1984. The glucosinolate content of some fodder brassicas. J. Sci. Food Agric. 35, 977-981.

Craig, E.A. and Draper, S.R. 1979. The glucosinolate content of oilseed rape varieties. J. Natn. Inst. Agric. Bot. 15, 98-103.

Carlson, D.G., Daxenbichler, M.E., VanEtten, C.H., Tookey, H.L. and Williams, P.H. 1982. Glucosinolates in crucifer crops. Turnips and rutabagas. J. Agr. Fd Chem. 29, 1235-1239.

Daun, J.K. & McGregor, D.I. 1983. Glucosinolate analysis of rapeseed (Canola). Method of the Canadian Grain Commission Grain Research Laboratory (Revised September 1983).

Daxenbichler, M.E., VanEtten, C.H. and Williams, P.H. 1979. Glucosinolates and derived products in cruciferous vegetables: analyses of fourteen varieties of Chinese cabbage. J. Agric. Fd Chem. 27, 34-37.

E.E.C. 1985. Annex VII [to regulation (FEC) No. 1470/68].

Fenwick, G.R., Spinks, E.., Wilkinson, A.P., Heaney, R.K. and Legoy, M.A. 1986. Effect of processing on the antinutrient content of rapeseed. J. Sci. Food agric. 37, 735-741.

Heaney, R.K. and Fenwick, G.R. 1980a. The analysis of glucosinolates in Brassica species using gas chromatography. Direct determination of the thiocyanate ion precursors, glucobrassicin and neoglucobrassicin. J. Sci. Food Agric., 31, 593-599.

Heaney, R.K. and Fenwick, G.R. 1980b. Glucosinolates in Brassica vegetables. Analysis of 22 varieties of Brussels sprout (Brassica oleracea var. gemmifera). J. Sci. Food Agric. 31, 785-793.

Heaney, R.K. and Fenwick, G.R. 1981. A micro-column method for the rapid determination of total glucosinolate content of cruciferous material. Z. Pflanzenzucht. 87, 89-95.

Helboe, P., Olsen, O. and Sørenson, H. 1980. Separation of glucosinolates by high performance liquid chromatography. J. Chromatography, 197, 199-205.

Lein, K.A. 1970. Quantitative Bestimmungsmethoden fur Samenglucosinolate im Brassicaarten und ihre Anwendung in die Zuchtung von Glucosinolatarmen Raps. Z. Pflanzenzucht. 63, 137-154.

Lewis, J. and Fenwick, G.R. 1986. Glucosinolate content of Brassica vegetables: Analysis of twenty four cultivars of calabrese (green sprouting broccoli, Brassica oleracea L. var. botrytis subvar. cymosa Lam.) in preparation.

Lila, M. and Furstoss, V. 1986. Choice of specific wavelengths for glucosinolate analysis in whole rapeseed by near infrared reflectance spectroscopy. Proceedings NIRS Symposium, Scheveningen.

Luthy, J., Carden, B., Bachmann, M., Friderich, U. and Schlatter, C.H. 1984. Nitrosierbare Stoffe in Lebensmitteln: Identifizierung und Mutagenitat des Reaktionsproduktes von Goitrin und Nitrit. Mitt. Gebiete Lebensm. Hyg. 75, 101-109.

Maheshwari, P.N., Stanley, D.W., Gray, J.I. and Van de Voort, F.R. 1979. An HPLC method for the simultaneous quantitation of individual isothiocyanates and oxazolidinethione in myrosinase digests of rapeseed meal. J.A.O.C.S. 56, 837-841.

McGregor, D.I., Mullin, W.J. and Fenwick, G.R. 1983. Analytical methodology for determining glucosinolate composition and content. J. Assoc. Off. Anal. Chem. 66, 825-849.

McGregor, D.I. 1985. Determination of glucosinolates in brassica seed. Eucarpia-Cruciferae Newsletter No. 10, 132-136.

McDanell, R., McLean, A.E.M., Hanley, A.B., Heaney, R.K. and Fenwick, G.R. 1986. Differential induction of mixed function oxidase (MFO) activity: a correlation with the levels of intact glucosinolates and glucosinolate hydrolysis production in the cabbage . Food Chem. Toxicol.

McMillan, M., Spinks, E.A. and Fenwick, G.R. 1986. Preliminary observations on the effect of dietary Brussels sprouts on thyroid function. Human Toxicol. 5, 15-19.

Mellon, F. and Spinks, E.A. 1986. (unpublished results).

Minchinton, I., Sang, J., Burke, D. and Truscott, R.J.W. 1982. Separation of desulphoglucosinolates by reverse-phase high performance liquid chromatography. J. Chromatography 247, 141-148.

Møller, P., Plöger, A. and Sørensen, H. 1985. Quantitative analysis of total glucosinolate content in concentrated extracts from double-low rapeseed by the Pd-glucosinolate complex method. In: (ed. Sørensen, H.) Advances in the production and utilization of cruciferous crops, Nijhoff/Junk; Dordrecht, Boston and Lancaster, pp 97-110.

Sang, J.P. and Truscott, R.J.W. 1984. Liquid chromatographic determination of glucosinolates in rapeseed as desulphoglucosinolates. J. Assoc. Off. Anal. Chem. 67, 829-833.

Sones, K., Heaney, R.K. and Fenwick, G.R. 1984a. The glucosinolate content of U.K. vegetables - cabbage (Brassica oleracea), swede (B. napus) and turnip (B. campestris). Fd Add. Contam. 1, 289-296.

Sones, K., Heaney, R.K. and Fenwick, G.R. 1984b. Analysis of twenty-seven cauliflower cultivars (Brassica oleracea L. var. botrytis subvar. cauliflora D.C.). J. Sci. Food Agric. 35, 762-766.

Sørensen, H. 1985. Limitations and possibilities of different methods suitable to quantitative analysis of glucosinolates occurring in double low rapeseed and products thereof. In: (ed. Sørensen, H.) Advances in the production and utilization of cruciferous crops. Nijhoff/Junk; Dordrecht, Boston and Lancaster pp. 73-84.

Spinks, A.E., Sones, K. and Fenwick, G.R. 1984. The quantitative analysis of glucosinolates in cruciferous vegetables, oil seeds and forage crops using high performance liquid chromatography. Fette, Seifen, Anstrichm. 86, 228-231.

Starr, C., Suttle, J., Morgan, A.G. and Smith, D.B. 1985. A comparison of sample preparation and calibration techniques for the estimation of nitrogen, oil and glucosinolate content of rapeseed by near infrared spectroscopy. J. Agric. Soc., Camb. 104, 317-323.

Thies, W. 1976. Quantitative gas liquid chromatography of glucosinolates on a microliter scale. Fette, Seifen, Anstrichm. 78, 231-234.

Thies, W. 1977. Analysis of glucosinolates in seeds of rapeseed (Brassica napus L.). Concentration of glucosinolates by ion exchange. Z. Pflanzenzuchtg, 79, 331-335.

Thies, W. 1979. Detection and utilisation of a glucosinolate sulfohydrolase in the edible snail, Helix Pomatia. Naturwissenschaften, 66, S. 364.

Thies, W. 1982. Complex-formation between glucosinolates and tetrachloropalladate (II) and its utilization in plant breeding. Fette, Seifen, Anstrichmittel 84, 338-342.

Thies, W. 1985. Determination of the glucosinolate content in commercial rapeseed loads with a pocket reflectometer. Fette, Seifen, Anstrichmittel 87, 347-350.

Underhill, E.W. and Kirkland, D.F. 1971. Gas chromatography of trimethylsilyl derivatives of glucosinolates. J. Chromatography 57, 47-54.

VanEtten, C.H., McGrew, C.E. and Daxenbichler, M.E. 1974. Glucosinolate determination in cruciferous seeds and meals by measurement of enzymically released glucose. J. Agric. Food Chem. 22, 483-487.

VanEtten, C.H., Daxenbichler, M.E., Williams, P.H. and Kwolek, W.F. 1976. Glucosinolates and derived products in cruciferous vegetables: Analysis of the edible part from twenty-two varieties of cabbage. J. Agric. Fd. Chem. 24, 452-455.

Vermorel, M., Heaney, R.K. and Fenwick, G.R. 1986. Nutritive value of rapeseed meal: Effects of individual glucosinolates. J. Sci. Fd Agric. (in press).

Wakabayashi, K., Nagao, M., Tahira, T., Saito, H., Katayama, M., Marumo, S. and Sugimura, T. 1985. 1-Nitrosoindole-3-acetonitrile, a mutagen produced by nitrile treatment of indole-3-acetonitrile. Proc. Japan Acad., 61, 199.

Wattenberg, L.W. and Loub, W.D. 1978. Inhibition of polycyclic aromatic hydrocarbon induced neoplasia by naturally occurring indoles. Cancer Research 34, 1410.

Youngs, C.G. and Wetter, L.R. 1967. Microdetermination of major individual isothiocyanates and oxazolidinethione in rapeseed. J.A.O.C.S. 44, 551-554.

CONCLUSIONS AND RECOMMENDATIONS

The lectures were followed by a general discussion and a restricted meeting.

GENERAL DISCUSSION

This session was chaired by Prof. Dr. Robbelen who pointed out that although presentations were not yet all complete; the work of the six Contractors was still sponsored until the end of the year.

There was a lengthy discussion about the types of methods available, the suitability of rapid methods and about the point at which such methods should be applied.

Delegates agreed that there is an urgent need for fast and accurate methods.

The suggestion that monitoring the certified seed supplied to the farmer was not supported by Dr. Aitzetmuller and Dr. Sørensen.

The German view was that methods could be divided roughly into 3 types :

A Crude "hammer test" type

B E.C. methods to determine payment

C HPLC / GLC type

It was further suggested that method B should be quick and reliable and that HPLC type methods C should be used as a check. It was further argued (Dr. Aitzetmuller) that the E.C. method should be applied to all samples.

Delegates were asked to submit an account of their experiences with particular rapid methods. It was agreed that Prof. Dr. Robbelen, Drs. Biston, Sørensen and Drs. Mc Gregor and Muuse would each submit accounts of their experience with "hammer test", N.I.R., palladium test, glucose test and thymol/sulphuric test.

Mr. Connell (C.E.C.) expressed concern that a suggestion for a suitable method for Community trade seemed to be missing and he asked if the six Contractors had produced a reliable method.

Prof. Dr. Robbelen said that it was hoped to present a method at the end of the period but added that easier methods were needed.

RESTRICTED MEETING OF THE SIX CONTRACTORS

The restricted meeting was chaired by Prof. Dr. Robbelen and in addition to the Contractors, Drs. Mc Gregor and Uppstrom were invited to attend.

The meeting was confined to discussion in some detail of what might eventually be proposed as an Official Method in the Community.

Although it had previously been agreed among some of the Experts that both the HPLC of intact glucosinolates with ion-pairing and the HPLC of desulphoglucosinolates might be appropriate, Mr. Connell said that a "two methods" solution was unacceptable to the Commission at this stage.

The justification for such an approach was that the possible introduction into oilseed rape of unusual glucosinolates (coumaroyl compounds etc) through breeding programmes can only be measured by analysing intact glucosinolates.

Currently, only new breeding lines would need to be monitored for such compounds. If and when they occured in commercial seed, any Official Method could be revised.

Accordingly since several Scientists were already familiar with-and using the desulphoglucosinolate method, this was tentatively proposed as the basis for an Official Method.

The Chairman prompted discussion of each stage of the different analytical processis with the following conclusions .

Extraction and desulphation: optimized procedures based on research by Drs. Buchner, Ribaillier, Quinsac and Wathelet would be adopted and included in a revised protocol as defined by Dr. Sørensen paper.

Internal standard : the internal standard should be included during extraction and should be glucotropaeolin and preferably glucobarbarin (since some columns perform differently).

Dr. Wathelet argued that the internal standard must be a commercial compound with known purity.

Commercial supplies of these compounds are, at this time, non-existant but Dr. Mc Gregor offered to supply glucobarbarin at cost until a suitable European source could be guaranted. Dr. Sørensen indicated that he would sow some barbaris seed to provide the initial requirement. Mr. Heaney undertook to renew contact with Sigma Chemical Company about this problem.

HPLC oven temperature: the temperature of the oven containing the column must be well defined. A high temperature gives a loss in 4 OH glucobrassicin (Drs. Buchner and Wathelet).

Response factors :response factors are essential to the accuracy of the method, and the Group undertook to reach agreement on this matter and apply the results to control samples of rapeseed.

Ring Testing : an initial ring-test of eleven rapeseed samples was proposed. Triplicate analyses by the six laboratories must agree and the defined procedure must be followed exactly. Results and comments should be sent to Dr. Buchner at Göttingen.

A wider ring test organised "blind" would then be conducted and involve more laboratories.

Dr. J-P Wathelet

LIST OF PARTICIPANTS

AITZETMULLER K.	Federal Center for Lipid Research 76, Piusallee D-4400 Münster GERMANY
BAUDART E.	Faculté des Sciences Agronomiques de l'Etat Chimie Organique et Biologique B-5800 Gembloux BELGIUM
BISTON R.	Station de Haute Belgique 48, Rue de Serpont B-6600 Libramont BELGIUM
BJERG B.	Royal Veterinary and Agricultural University 40, Thorvaldsenvej DK-1871 Frederiksberg C DENMARK
BUCHNER R.	University Institute of Plant Breeding 8, Von Sieboldstr. D-3400 Göttingen GERMANY
CASIMIR J.	Faculté des Sciences Agronomiques de l'Etat Chimie Organique et Biologique B-5800 Gembloux BELGIUM
CONNELL J.	C.E.C. 86, Rue de la Loi B-1040 Bruxelles BELGIUM
CWIKOWSKI M.	Faculté des Sciences Agronomiques Chimie Générale et Organique B-5800 Gembloux BELGIUM
DARDENNE G.	Faculté des Sciences Agronomiques de l'Etat Chimie Organique et Biologique B-5800 Gembloux BELGIUM
DE HARO Y BAILLON A.	S.I.A. Junta Andalucia APDO 240 Cordoba SPAIN
DOITSINIS A.	Cotton and Industrial Plant Institute Sindos Thessaloniki GREECE

DOMINGUEZ J.	S.I.A. Junta Andalucia APDO 240 Cordoba SPAIN
EGGUM B.	National Institute of Animal Science Foulom Research Centre DK-8833 Ørum Sønderlyng DENMARK
FROHLICH A.	The Agricultural Institute Oak Park Carlow IRELAND
GAZAGNES J-M	C.E.C. 200, Rue de la Loi B-1040 Bruxelles BELGIUM
HEANEY R.	Institute of Food Research Colney Lane NR4 7UA Norwich GREAT BRITAIN
LECOMTE R.	Centre de Recherches Agronomiques de l'Etat 22, Avenue de la Faculté B-5800 Gembloux BELGIUM
LEONI O.	Istituto Sperimentale per le colture Industriali 133, Via di Corticella 40129 Bologna ITALY
LOGNAY G.	Faculté des Sciences Agronomiques de l'Etat Chimie Générale et Organique B-5800 Gembloux BELGIUM
MARLIER M.	Faculté des Sciences Agronomiques de l'Etat Chimie Générale et Organique B-5800 Gembloux BELGIUM
MASSAUX F.	Faculté des Sciences Agronomiques de l'Etat 1, Passage des Déportés B-5800 Gembloux BELGIUM
Mc GREGOR D.	Agriculture Canada Research Branch 107, Science Crescent, Saskatoon Saskatchewan S7N OX2 CANADA

MEYERS R.	Services Techniques de l'Agriculture Av. Salentiny BP 75 Ettelbrück LUXEMBURG
MUUSE B.	Rikilt 45, Bornsesteeg Wageningen HOLLAND
PALMIERI S.	Istituto Sperimentale per le colture Industriali 133, Via di Corticella 40129 Bologna ITALY
PARSONS D.	Plant Breeding Institute Maris Lane, Trumpington Cambridge CB22LQ GREAT BRITAIN
QUINSAC A.	Cetiom Avenue de la Pomme de Pin, Ardon F 45160 Olivet FRANCE
RENARD M.	I.N.R.A. Station d'Amélioration des Plantes BP 29 F 35650 Le Rheu FRANCE
RIBAILLIER D.	Cetiom Avenue de la Pomme de Pin, Ardon F 45160 Olivet FRANCE
ROBBELEN G.	University Institute of Plant Breeding 8, Von Sieboldstr. D-3400 Göttingen GERMANY
SEVERIN M.	Faculté des Sciences Agronomiques de l'Etat Chimie Générale et Organique B-5800 Gembloux BELGIUM
SMITHARD R.	University of Newcastle upon Tyne Newcastle upon Tyne NE 4 9NX GREAT BRITAIN
SØRENSEN H.	Royal Veterinary and Agricultural University 40, Thorvaldsenvej DK-1871 Frederiksberg C DENMARK

UPPSTROM B.

Svalöf AB
S-268 00
Svalöf
SWEDEN

VAN DER MEER J.

Institute of livestock feeding and nutrition research
PO Box 160
8200 AD Lelystad
HOLLAND

VAN HEE L.

Rukstation Voor Plantenveredeling
Van Gansberghelaan, 92
B-9220 Merelbeke
BELGIUM

WAGSTAFFE P.

C.E.C.
200, Rue de la Loi
B-1049 Bruxelles
BELGIUM

WATHELET B.

Faculté des Sciences Agronomiques de l'Etat
Chimie Organique et Biologique
B-5800 Gembloux
BELGIUM

WATHELET J-P

Faculté des Sciences Agronomiques de l'Etat
Chimie Générale et Organique
B-5800 Gembloux
BELGIUM

WAUTERS M.

C.E.C.
86, Rue de la Loi
B-1040 Bruxelles
BELGIUM

GPSR Compliance

The European Union's (EU) General Product Safety Regulation (GPSR) is a set of rules that requires consumer products to be safe and our obligations to ensure this.

If you have any concerns about our products, you can contact us on ProductSafety@springernature.com

In case Publisher is established outside the EU, the EU authorized representative is:

Springer Nature Customer Service Center GmbH
Europaplatz 3
69115 Heidelberg, Germany

Batch number: 10531666

Printed by Printforce, the Netherlands